The FreeDOS Kernel

Pat Villani

Routledge
Taylor & Francis Group

LONDON AND NEW YORK

First published 1996 by CMP Books

This edition published 2015 by Focal Press

Published 2021 by Routledge
2 Park Square, Milton Park, Abingdon, Oxon OX14 4RN
605 Third Avenue, New York, NY 10017

Routledge is an imprint of the Taylor & Francis Group, an informa business

ISBN 13: 978-0-87930-436-2 (pbk)

To Bill, who tried to get me to write a book for 15 years.

To Kiddle, Mike, and Mom, for nodding and smiling.

And especially to Donna, whose love and confidence in me made this book possible.

Table of Contents

Chapter 1

Introduction

Ask anyone to define an operating system. I will bet that for every question you ask, you will receive a different answer. You may ask, "Is the definition of an operating system so vague?" I'm happy to answer — no. However, operating systems do mean different things to different people.

To many computer users, it is the full collection of utilities and the kernel that drives their computer to run their favorite applications. To engineers working with small microprocessor systems, it may be the kernel and drivers together that makes a complex control system work. No matter what the definition, an operating system of one form or another is an integral part of software engineers' and users' lives.

Also, operating systems are the subject of heated debates and argument. The debates range from the OS/2 versus Windows battles to the Berkeley versus System V UNIX debates. Lately, with the introduction of WindowsNT, the WindowsNT versus UNIX debate dominates discussions in many a workplace. These discussions typically result in heated debates that approach religious fervor over the choice of an operating system.

A Brief History of Desktop Operating Systems

Historically in the minicomputer and mainframe world, there were as many operating systems as there were computer vendors. Each vendor would design a unique operating system that took advantage of its unique hardware. Sometimes, the operating system changed drastically from one revision to another as the hardware changed, requiring major rewrites of application code. Fortunately, this was not the case for the desktop computing environment.

With the introduction of microprocessors, many individuals saw an opportunity to create small desktop computers. An 8-bit microprocessor, such as the Intel 8080 or Motorola 6800, was the central processing unit for the early desktop computers. Memory limitations were on the order of a few kilobytes, and some rather amazing applications ran in this small space. As the popularity of these computers grew, so did a small cottage industry serving hobbyists and small businesses.

Early users of these computers recognized advantages in sharing technology, and some early computer conferences attempted to standardize media formats and storage methods so programs be could exchanged. The storage medium for early desktop computers was audio cassette, with recording formats such as Tarbell and Kansas City Standard. Later, 8-inch and 51/4-inch floppy disks became affordable, and small operating systems such as CP/M and Flex further promoted standardization while still allowing room for experimentation.

The market for the desktop computer grew rapidly as did the complexity of microprocessors. Processors such as the Intel 8086 and the Motorola 68000 began promoting 16-bit and 32-bit capabilities, and companies such as Apple and IBM took advantage of these new processors. Operating systems developed for these new systems spawned a new generation of applications. These companies also spent considerably more time and money in planning markets than the earlier cottage industry that sprung up around desktop computers. Of particular interest to us is a case study of the joint IBM and Microsoft approach with MS-DOS. Its evolution will help shed some light on features and functions within MS-DOS.

MS-DOS

IBM's entry into the desktop market was with the IBM-PC. This desktop computer was the product of a skunk-works team within IBM that featured an Intel 8088 microprocessor as its CPU and 64Kb of RAM. Originally introduced with only cassette tape as standard storage, the IBM-PC also offered a 51/4-inch floppy disk as an option. To manage the diskette, IBM licensed the MS-DOS operating system from Microsoft.

The original design of MS-DOS evolved from an operating system developed by a Seattle company for its S-100-based desktop computers. Microsoft purchased all rights to it, because operating systems were new to them and IBM had approached them for software to run on its new PC. This system became the source code base for MS-DOS.

The MS-DOS precursor was similar to the CP/M operating system for the Intel 8080 and required little RAM, a distinct advantage because the 80x86 family is similar in design to the 8080. By design, the operating system provided a similar API (Application Programming Interface) to CP/M. As a result, MS-DOS 1.0 held very closely to the CP/M model. When combined with a familiar architecture, many developers were able to adapt their CP/M-based applications to MS-DOS. Favor quickly shifted toward Microsoft to the detriment of Digital Research, whose CP/M-86 operating system was the primary competition.

As the market matured, so did both MS-DOS and the PC. Demands for MS-DOS began to grow from outside IBM, with IBM leading the pack in technology. The next generation of desktop computers from IBM was the XT. This computer was similar to the PC but featured more memory and a 10Mb internal disk drive. Its motherboard also was capable of much more memory: 256Kb instead of 64Kb. IBM's target for this more powerful system was the business market. Previously, only large businesses could afford any kind of computing power. The XT addressed the needs of smaller businesses.

In order to address this new market, Microsoft developed a new strategy for their products as well. They adopted a dual approach to desktop computing environments. To manage the new internal disk drive, Microsoft adapted its XENIX operating system (an AT&T UNIX v7 derivative) for the XT. They also enhanced the MS-DOS operating system to support larger storage devices and added a XENIX-like hierarchical file system. Microsoft developers enhanced the MS-DOS API with many XENIX-like calls, which enabled easy migration of applications developed for a personal computer market based on MS-DOS to a business market based on XENIX. Microsoft also enhanced the MS-DOS command line interpreter, command.com, with I/O redirection and pipes similar to XENIX. This redesigned MS-DOS now became the basis for all new MS-DOS versions.

The lock-step product announcements continued with both IBM and Microsoft again adding new products in unison. IBM introduced the AT, based on an Intel 80286, and Microsoft released another version of MS-DOS to support it. At this time, Microsoft began to abandon its earlier XENIX product in favor of LAN connectivity and GUI interfaces. IBM also started a joint development with Microsoft for OS/2. This new operating system would take advantage of memory management and other advanced features that the new processor offered. The need for a new operating system was clear — MS-DOS lacked protection mechanisms, and it was very easy for a program crash to destroy the operating system. The Ctrl-Alt-Delete key sequence became very familiar to users, and a huge market for fixed-disk reconstruction programs was born.

For migration of MS-DOS applications to OS/2, IBM and Microsoft developed a strategy similar to their market entry strategy. They would provide all the functions that were previously available so that developers could quickly adapt existing applications to the new operating system. Virtual devices and API entries similar in functionality to BIOS calls eliminated the need for int xx-style real-mode device control, while retaining the functionality familiar to developers. With API calls defined in C, OS/2 encouraged greater developer productivity through the use of high-level languages and broke the traditional register call model that both CP/M and MS-DOS sported. Unfortunately, a design limitation of the 80286 forced some serious incompatibilities between OS/2 and MS-DOS.

The Intel 80286 featured two modes of operation, a real mode and a protected mode. The 80286 protected mode sported a new memory management scheme that modified the processor behavior so that memory access is very different from real mode. Any access to memory uses the contents of descriptors, in place of segment registers, to look up a new base address and add it to the offset address. The descriptor plus offset scheme expanded memory capacity from 1Mb in the 8086/8088 to 16Mb in the 80286. An unfortunate side effect of this scheme is that real-mode programs using segment plus offset addressing will not port easily to protected mode.

Additionally, a significant design oversight on Intel's part existed in the 80286. The processor could switch from real mode to protected mode in order to take advantage of features offered by the new operating system, but it could not switch back. Once in protected mode, the only way back was a hardware reset to the 80286. IBM designed new features into both the BIOS and hardware of the AT to make the switch back to real mode, but the context switch time was very high. Thus, new multitasking OS/2 applications worked in protected mode only, and MS-DOS real-mode applications were limited to a single task.

The OS/2 protected-mode environment was similar to environments found on larger computers but was incompatible with existing MS-DOS applications. This limited its popularity — users did not want to drop all their old applications and spend much more for special protected-mode applications — and generated bad press about OS/2. When combined with rumors of discord between the successful IBM/Microsoft team, users stayed with MS-DOS. Microsoft decided at this point to create a migration path that would always include MS-DOS compatibility for all its new operating systems.

New directions from Microsoft included a new GUI environment, Windows, that layered MS-DOS. For all subsequent releases after 3.3, MS-DOS and Windows became more tightly coupled so that both could cooperatively take advantage of 80386+ processors. These new processors featured improved memory management, linear addressing, and a V86 mode. The V86 mode allowed real-mode programs to function while taking advantage of operating system services for process protection and multitasking. Through it all, MS-DOS remained the foundation for 80x86 desktop computing. This may change in the future, now that early machines are being retired from service, and Microsoft, feeling relief from supporting the older real-mode only computers, has many new, advanced operating systems in the wings.

DOS-C

DOS-C started as an experiment in writing device drivers using C for MS-DOS v3.1. Many articles and books that discussed the foundation of a DOS device driver had appeared on the scene; however, it seemed that each article described code written in assembly language only. In a discussion with some colleagues, I argued that this was unnecessary, citing UNIX as an example. As a result of these discussions, I began to write DOS device drivers in C. I developed both block and character device drivers, along with special C data structures to match the DOS request packet. This effort proved to be very successful, and I decided to continue with personal research to expand on these techniques. The goal: to create a full operating system by using the same techniques used to create the device drivers.

I began the design of a new architecture that would use assembly language only at the call interface and for process management, such as stack manipulation and context switching. The implementation of this operating system would take advantage of the C language features and require fewer resources to develop than the traditional assembly language designs. Although UNIX proved this concept for large multi-user operating systems, there were no documented instances for a traditional PC operating system.

Building on this knowledge, I developed a minimal operating system using the device drivers written earlier, along with a new 8086 interrupt API. Known as XDOS, it proved to be a functional operating system. This new operating system became the vehicle to explore booting techniques, and I developed a C library SDK (Software Development Kit) for it.

I later began to enhance XDOS and chose DOS as the new API. These enhancements included a more advanced architecture that uses an IPL (Intermediate Program Loader) to set up the operating environment before loading the operating system itself. Additionally, reentrant system calls for real-time applications became part of the design. Using the name NSS-DOS (bearing the initials of the consulting firm I was part of in place of Microsoft's MS), I demonstrated this version to a few friends and business acquaintances. As a result of these demonstrations, a major defense contractor approached my consulting firm for a source license for this operating system. The only new requirement — it had to run on 680x0 processors.

This presented a new challenge. Because of the MS-DOS model used for the API, NSS-DOS relied heavily on a segmented architecture. To meet this challenge, I began a major redesign of NSS-DOS. Development of new portability techniques allowed compilation of the same source code base on a variety of hosts and with a wide range of compilers and target processors. The new version, DOS/NT, resulted from this work. The redesigned kernel used microkernel techniques with logical separation of the file system, memory, and task managers. The new design included a portable DOS API along with a new DOS SDK to guarantee portability. By removing all processor-unique code from the core functions, a highly portable operating system resulted.

The version described in this book, DOS-C, is derived from DOS/NT and is closer in design to MS-DOS: the kernel is one large program as opposed to many small programs. It is a nonmultitasking operating system that provides a large number of system calls similar to MS-DOS. This simplified design provides an operating system that readers can study without getting lost in the complexity that a microkernel entails. It also provides an excellent source code base to experiment on or with.

Why Roll Your Own

Although I discussed the reasons for the development of DOS-C, there are probably many readers who are wondering why bother at all. After all, a professional software company with a great deal of resources behind it wrote and maintains MS-DOS. Why would anyone develop a simple clone of a successful operating system?

Educational Tool

The first application for such an operating system is as an educational tool. Without a collection of reverse engineering tools, it is impossible to examine MS-DOS in any great detail. Anyone who has ever looked at maintenance of a software system will tell you that trying to get into the original programmer's head to figure out the reasons for a particular design, without proper documentation, is a monumental task — especially at the disassembled assembly code level. My compliments to all the folks who are pouring out reams of "undocumented documentation." Reverse engineering MS-DOS may also break any existing licensing agreements between you and Microsoft, so check your license before applying a disassembler to it.

My guess is that MS-DOS source code is almost entirely in assembly language, and the kernel itself is guaranteed to be assembly language. Even if you did see the source, it would be more difficult to follow than if coded in a high-level language. My goal of minimizing assembly language yields two benefits: the code is portable and easier to understand. Although a high-level language in itself does not necessarily guarantee portability, it is easier to port C source code than assembly source code. Also, conditional code and looping constructs are easier to understand in C than assembly language.

As an educational tool, you can replace any section of the tested operating system. This allows you to experiment with different algorithms. You may want to change the LRU (Least Recently Used) buffer algorithm for a different one and compare performance. You may also want to learn about the different algorithms in order to apply a similar technique to a different application. For example, loadable device drivers may be useful in a stand-alone application where the host uses different devices in different configurations, or the buffered keyboard input code may provide enough insight for an interactive display system.

Finally, you can test the code on a host MS-DOS or UNIX machine using a source-level debugger because the majority of DOS-C is C code. For example, performing DOS/NT code tests on a 680x0-based UNIX machine proved invaluable. Many of the "big-endian" versus "little-endian" issues surfaced quickly during the unit testing of individual components of the operating system.

Embedded Systems

Possibly one of the most useful applications is as an embedded operating system. Many times, an application requires the functionality of an operating system, such as file storage, embedded databases, or sophisticated device control. A general-purpose operating system used in an embedded system may not do the trick. Additionally, it may not be possible to develop the application on the host because of a lack of hardware or support tools.

You can simplify the design of a typical embedded system by using a DOS PC for application development and later linking to your DOS to create an application ROM. You can debug the application on the host PC and define all the DOS entry points used. Later, you can either replace the DOS functionality or use an embedded operating system with custom device drivers for your unique hardware. The embedded operating system approach eliminates work and becomes a tool for the developer. Additionally, the embedded operating system becomes an abstraction layer between the application and the embedded system hardware.

Non-Intel Applications

When the operating system design is portable, new opportunities open up for both traditional and embedded applications. One possibility in the area of embedded systems is to design your system and develop it as you would for a normal PC embedded system, then use cross-development tools to recompile the application. When combined with the cross-compiled operating system, software verification reduces to the level of regression testing on the new hardware.

Another reason for developing your own operating system is simply to adapt it to a new platform. There are many non-Intel processor single-board computers. To approach a company such as Microsoft to adapt MS-DOS to this custom hardware would be both time and cost prohibitive. However, with an operating system designed to be portable, moving the operating system to a new target is a question of cross-development tools.

DOS Clone

Another way to view DOS-C is as a declaration of independence. As the desktop computer market matures, Microsoft is moving away from MS-DOS as the foundation of its GUI products. Another example of the trend away from DOS is the decreasing size of the DOS clone market. Products such as DR-DOS are rapidly disappearing from the scene, which leaves older applications in an orphan state. Creating your own version of DOS solves this problem.

Extending DOS

Have you, as a developer, ever looked at the DOS API and felt that if it had just one additional feature, your professional life would be easier? Taking the challenge head-on and creating your own version of DOS is an excellent way of extending DOS.

A typical way to extend DOS is with the Novell approach. When they created their network product, it was as an extension to the existing MS-DOS `int 21h` call. They captured the call and searched their own system calls. If it matched one, the network product would service it. Otherwise, the processor was restored to its original state and the call was passed through to DOS. Microsoft also adopted this interrupt chaining with its standard `int 2fh` multiplex interrupt chain in v3 and above.

With your own DOS, you can extend system calls to cover your own special call by simply adding it to the function call handler. You write an extension, test it, and integrate it into your DOS. The simplicity behind this type of customization gives you, the developer, enormous power over your applications environment.

Development Environment

I expect to hear you now say: "Gee, this all sounds great, but what expensive tools do I need to do this?" As it turns out, you do not necessarily need any. By using a native 80x86 platform to develop for and test on, popular PC compilers from companies such as Borland and Microsoft are suitable as long as you follow special guidelines. Also, there is no need for special linkers if you design your architecture correctly. As I will describe in later chapters, the use of an IPL simplifies the files on disk, and you can use common .com and .exe file formats. Testing also becomes simpler through the use of some low-cost tools. With all this in mind, choose your tools.

Compiler

First, choose a C compiler. The version of DOS presented in this book compiles with Borland C (Borland International, Scotts Valley, Calif.), Microsoft C (Microsoft Corp., Redmond, Wash.) and other C cross-compilers. Special attention to the use of portable techniques pays off when using simple development tools. As it turns out, the only area that ties a given C compiler to its host operating system is the C runtime library. Avoid its use to ensure that your code will run stand-alone with the low-cost compilers.

Beware of floating point code used by the C compiler. Some compilers generate non-reentrant floating point code or code that uses non-reentrant library calls. Many times, a compiler's start-up code will take over a number of interrupt vectors. These vectors capture floating point exceptions generated when a call to a nonexistent coprocessor occurs. This is a common technique for automatic switching between coprocessor floating point calculations and software emulation of the coprocessor. For all these reasons, and more, avoidance of floating point code helps. DOS-C avoids all floating point code by avoiding the use of C library calls such as printf(), further simplifying the choice.

Assembler

Besides the C compiler, you will need an assembler. The need for this is twofold. First, the standard start-up code delivered with a native C compiler assumes a host machine and operating system. You will be working with a "naked" machine that does not have an operating system available (remember, you are creating it). As a result, you need to substitute your own start-up code that will take care of memory initialization, stack and register initializations, etc., and finally call main(). Second, you will use the de facto DOS standard. This means that you use the same registers for system calls, switch stacks during system calls, etc.

The support code must be written in assembly language, allowing us full control of the target machine. The asm shortcut provided by many C compilers simply does not suffice because we must move away from the C model for this support. You cannot afford stack register and frame pointer manipulations, and many times you will not generate code that resembles a C function.

Linker

Look at the linker provided with the C compiler you choose. For our 80x86 version, you can use the standard DOS linkers. However, for many platforms they may not be suitable. You may need some form of "strip" utility as typically supplied on many UNIX platforms so that you can adequately create an IPL. As you will see later, our architecture specifies that the kernel is an executable file for a known format. Our choice of the .exe file type simplified our decision, but you will need to be careful when examining other platforms and targets.

Locator

If you have decided to place the operating system into ROM, or if you have a special target machine architecture that requires you to fix the operating system in memory, you may also need a locator. This type of utility comes in various shapes and sizes and may be split between the linker and various other utilities, but it always serves the same purpose: to fix all memory references to absolute addresses and possibly to generate a special file format acceptable to various device programmers. There are two examples of this type of utility set. The first example is as a stand-alone locator. Multiple vendors in the 80x86 community have such utilities to convert DOS .exe files into "hex" files. UNIX users may find that their linker, ld, accepts special specification files and that other utilities will convert the resulting file into a hex file. Check your manuals because this is a function of the vendor whose UNIX you are running.

Debugger

With all your development tools in place, you will need to begin thinking about debugging tools. This choice is dictated by a number of factors. The most significant is your target architecture. If you plan to use your operating system in ROM, then a tool such as a ROM emulator that will accept a download from your development platform and emulate your target ROM(s) may be a good choice. This device will permit you to download your code directly, but you will need some other tool in combination with the ROM emulator to assist you in debugging.

If your budget permits, you may want to look into an in-circuit emulator. This tool replaces the processor on your target hardware and allows you to electronically look into your target. They typically provide all the favorite debugging methods available on host debuggers, such as single-step and break point. Additionally, they usually allow you to stop a processor that has stopped responding — a highly desirable feature for anyone who has experienced such a lock-up while running DOS-based native debuggers. Also, many other hardware features, such as trace buffers, are usually available for in-circuit emulators. This type of buffer allows you to look into the history of your program so that if you wonder how you got to a certain point in your program, the trace buffer will point it out. Finally, if source-level debugging is an option, you should buy it. It will simplify your job by orders of magnitude. Just make certain that it supports your development tools.

If you are on a limited budget or you are using DOS-C as a learning tool, you will want to look into remote debuggers. Many of you are probably already familiar with the remote debugging features of debuggers such as Turbo Debugger (Borland International) or CodeView (Microsoft Corp.). These run on a remote DOS machine, but the type of remote debugger you will want to use is one that runs stand-alone. They sometimes require some work to adapt to the target hardware but are well worth the effort. Some will also work in combination with ROM emulators, making the combined tools almost as powerful as an in-circuit emulator.

Getting Started

Now that I have briefly touched on why and how to write your own operating system, I will begin to build DOS-C. I will approach its development in a logical fashion by first developing a set of requirements. I will examine MS-DOS from the outside in order to assist in the development of a set of general specifications. I will develop a general-purpose method of examining error recovery and apply it to round out our requirements. When completed, I will proceed to develop the operating system itself.

The full source code for this book is now available on the publisher's ftp site at `ftp.mfi.com/pub/rdbooks/FreeDOS.zip`; login as "anonymous" and download the file. Any references to the "companion code diskette" in the book now refer to the code available on the ftp site. See Appendix C for more details or check the FreeDOS home page at `http://sunsite.unc.edu/pub/micro/pc-stuff/freedos` or my personal web page at `http:/www.monmouth.com/user_pages/patv` for updates.

## Chapter 2

DOS Basics

Where to Start

Many programmers have come to take MS-DOS for granted. Its API has become a standard in the industry and is known by developers world-wide. Many books, published by Microsoft and others, describe the interface. Many other books contain information about writing applications for DOS, and others talk about how to extend it. Still others describe how to write extenders, device drivers, etc. Additionally, periodicals dedicate space to and books are published about the "undocumented" system calls and data structures.

Although Microsoft produced a great deal of programmer and user documentation for MS-DOS, there is plenty of undocumented information about MS-DOS internals. Proof of this is the sheer volume of "undocumented" books and articles published about MS-DOS. However, even with the undocumented sources, the target audience for this documentation is application programmers who want to emulate some of Microsoft's private programming tricks, not for programmers who want to design their own version of DOS. Very little documentation exists about MS-DOS internals. You may need to provide the same interface as MS-DOS and in some cases emulate MS-DOS internals in your design in order to achieve any level of binary compatibility.

Whenever possible, I followed reference articles and texts during the design. These references are very good and cover a good portion of what you need to know. However, as pointed out earlier, questions will come up where you will need to perform little experiments to find out how DOS behaves. One reason is that many times, a programmer's references cover a subset of error responses; however, you will need to cover other undocumented instances. Another reason is that you can gain insight into the DOS design by running a test program that exercises a system call.

Exercising system calls through varying conditions is one of the best ways to study DOS. Many times, you don't know what algorithm the MS-DOS designers used. Quite often, it may be possible for many algorithms to perform certain functions. Unlike an open system where all functions of the operating system adhere to a well-known standard, Microsoft created the system functions and modified them as they went along. In some cases, the changes occurred as bug fixes or as the result of developer learning curves. In other cases, Microsoft created new system calls to take advantage of new features introduced with a release. As a result, many applications have become dependent on these features and any little quirks that they may exhibit. You need to emulate these features closely so that your version of DOS doesn't break existing applications. Although more than one algorithm may achieve a given function, you must choose the algorithms for your DOS that closely emulate the original MS-DOS version.

This chapter contains a sample experiment whose purpose is to examine the minimum set of files and other objects that MS-DOS requires to run. The goal is to demonstrate a methodology that you can use whenever a design question arises. From there, the chapter covers the basic DOS boot procedure, physical disk organization, API, file system, memory management, and device driver.

Basic MS-DOS Architecture

The way I like to study any new subject is by starting with the overall picture. I like to peel away the different layers, uncovering new information with each new layer. I find this approach helps me comprehend a new subject better because I will not get lost in the details too early in the game. I will take this approach and begin by looking at the four basic MS-DOS files: boot, `io.sys`, `msdos.sys`, and `command.com`. You may ask, "How do I know that these are the basic MS-DOS files?" Perform a simple experiment to find out.

You will format a bootable floppy disk as the basis of this experiment. Take a blank disk, place it into drive A, and type `format a: /s` at the DOS prompt. MS-DOS chugs along and gives you a bootable disk. You know it's bootable because you can now hit Ctrl-Alt-Delete and up comes MS-DOS. You can copy files, delete files, execute other programs, etc. In short, every essential MS-DOS function is available at the new prompt.

Now take a closer look at this bootable disk and see what MS-DOS has done to it, aside from laying out track and sector structure to make it bootable. When you examine the directory, you find one file, command.com, on it. This file is the MS-DOS command line interpreter or shell. If you do a dir a: /a:h, you find two hidden files, io.sys and msdos.sys (or ibmbio.sys and ibmdos.sys in certain versions of MS-DOS and PC-DOS). It seems that Microsoft considers these three files to be the minimum components needed for MS-DOS to operate.

This is interesting, but you know from the references that one other component will not appear in any directory listing because it is not a true file: the boot sector. The boot sector is the first sector placed onto the disk by the format command after the disk physical format operation. The PC architecture defines its function, and it is a de facto standard. All PC-style computers load and execute the boot sector in order to boot any operating system; it is the final component to the minimum set of programs for MS-DOS.

Each of the four components has a specific role in starting MS-DOS. As you can see, your little experiment proves that only four components are necessary for a minimal DOS operating system. If you limit your design to only these components (or their equivalent functions), your design becomes easier to handle by eliminating the development of transient commands such as attrib, format, etc.

Having established the minimum set of programs necessary for MS-DOS to operate, you can examine the dynamics of these four programs. If your computer is a standard PC-class computer like mine, the BIOS performs a power-on self test (POST) the instant you turn your computer on. The basic tests your computer may perform are a memory test, followed by a keyboard test, disk test, and hard drive test. You may also see coprocessor tests and other varied tests interspersed among the basic tests. At the end of these tests, you will hear a beep. The question now is, "What happens after the beep?"

The answer is simple: Your computer's BIOS now attempts to boot the operating system from either the floppy disk or the hard disk. If it's booting from a floppy, it brings the boot sector into memory at location 0:7c00h. This is an important address, because all BIOS guarantee to load the boot at this location for compatibility reasons. If the boot sector is not at 0:7c00h, MS-DOS will not boot properly because a closer examination of the boot program reveals that its address modes require this start address (unlike a .com or .exe file that can load anywhere within the x86 real mode memory). If it's booting from a hard drive, the BIOS loads a master boot sector to determine the active partition and then loads the boot record into the same address. However, for educational purposes, you will continue to examine a floppy disk boot.

After loading the boot, the BIOS performs a far jump to 0:7c00h. MS-DOS places a boot sector onto a disk or hard disk regardless of whether it is bootable or not. This protects the user from inadvertently attempting to boot from a nonbootable disk (unfortunately only for x86 processors). By design, the boot checks for a bootable hard or floppy disk by looking at the root directory for the presence of two required MS-DOS files, io.sys and msdos.sys. If boot does not find them, it issues the well-known "non-system diskette" message and waits for a new disk to retry. The hard drive boot does not wait but halts, since a boot failure means that something is seriously wrong with the active partition, whereas a floppy boot error may result from a user inserting the wrong floppy disk into the boot drive.

The boot next searches the root directory for io.sys and msdos.sys (or ibmbio.sys and ibmdos.sys). It reads each entry in the root directory in order to determine the filenames. If these files are present, boot proceeds to load io.sys. It does this by getting the starting disk sector from the matching io.sys entry, copying each sector into a fixed location in memory, and transfering execution control to it. One significant boot limitation is that it can only load sequential records. Boot's size limitation is significantly less than one sector, and the developers who created boot were very creative when it came to trading off functionality versus size.

Once the transfer of control takes place, `io.sys` proceeds to initialize its device drivers. These drivers are the fundamental device drivers that every DOS system has: floppy disk, hard drive, console, printer, and auxiliary serial port. Up until this point, boot communicated strictly through the BIOS functions in order to save space and achieve portability, but `io.sys` now uses its device drivers for control. `io.sys` continues to initialize its device drivers by loading `msdos.sys` and instructing it to initialize itself. It then uses `msdos.sys` to read `config.sys`, if present. `io.sys` parses each line of `config.sys` and sets up the environment space, loads and initializes the device drivers, and performs other functions necessary for a successful boot. When `io.sys` completes all this, it loads `command.com` and finally transfers control to it.

Notice that I have not mentioned the familiar DOS startup file, `autoexec.bat`. That is because it is really not part of the actual boot process but a convenience placed into `command.com`. The designers of MS-DOS knew that many people modified CP/M in order to execute a program on startup and that other operating systems, such as UNIX, had special files or directories of files that executed on startup (i.e., `/etc/rc`). These files are extremely useful in turnkey applications and tailoring environments. You can perform a short experiment to verify that `autoexec.bat` is a function of `command.com`. You can take the sample disk formatted earlier and add a simple `autoexec.bat` file to it. Next reboot your computer to verify that the new file executes. Then change `command.com` to a simpler or different command line interpreter. What you will see is the same as before, except without the execution of `autoexec.bat`. By verifying that `autoexec.bat` did not execute when the command line interpreter changed, you verified that its association is with `command.com`.

You now have enough data from this little experiment to generalize the MS-DOS boot procedure.

1. Four mandatory components of a basic MS-DOS system are: boot, io.sys, msdos.sys, and command.com. Two optional files are: config.sys and autoexec.bat.

2. To initiate a boot sequence, load the boot sector at 0:7c00h and execute it.

3. Boot must examine the disk. If bootable, it loads the first stage, io.sys (or ibmbio.sys), and executes it.

4. Upon transfer of control, io.sys loads msdos.sys (or ibmbio.sys loads ibmdos.sys) and allows it to initialize. io.sys then uses msdos.sys to parse and execute config.sys if present.

5. When completed, command.com is loaded and control is transferred to it.

6. command.com executes autoexec.bat if present and displays the familiar DOS prompt, awaiting user input for commands.

You will use this information later when you design your architecture for DOS-C.

What Do Physical Disks Look Like?

One area of mystery to many developers is the physical location of sectors stored on a disk. Unless you've developed disk drivers before, you've probably never seen this. Even if you did write a disk driver, odds are that you've never looked at the physical file system organization. I'll show you the physical and logical organization of the MS-DOS file system so that the mystery will finally be solved.

Generic Disk Organization

Disk drives are ubiquitous. Nearly every desktop computer, workstation, server, and mainframe has at least one. They are the preferred secondary storage for computers. Typically, their size is many times the primary RAM and ROM storage contained within the computer itself. They're also taken for granted. Why? Because storing data onto the disk is done without regard to either physical or logical organization. Just pause for one moment and ask yourself when was the last time you thought about which sectors the file you just created occupies?

Before I get into such details, you need to understand the terms used to describe how the disk stores data and how the disk actually works. Figure 2.1 is a view of a generic disk. This disk has one or more physical *platters* that hold the data. Each platter allows data to be read and written from either one or both surfaces due to a coating capable of being magnetized. This coating is similar to that used on mylar tapes

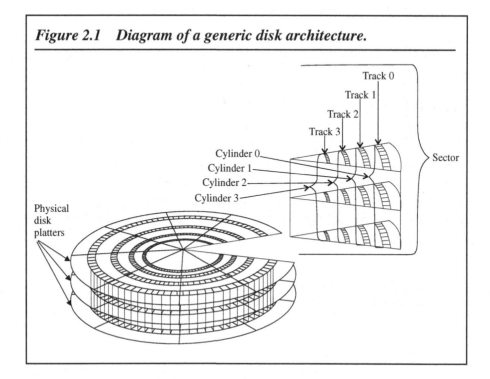

Figure 2.1   Diagram of a generic disk architecture.

that store data or audio. The data is stored by magnetizing a portion of this surface as the surface moves beneath a read/write head. Unlike a tape, where the tape is spooled from one reel to another, a disk rotates so that the same physical location is available once per revolution and data can be written and repeatedly accessed by waiting for the next revolution. This stripe of recorded data is called a *track*.

This track contains all the data. You need to access this data much the same way you access internal memory — you need to be able to modify it at random intervals. In order to facilitate this, the track is broken down into *sectors* that have prerecorded markers (laid down by a formatting program) that identifies where a sector is physically located on the track. By prior agreement, this sector is defined as a fixed number of bytes designed to fit within the physical space between the prerecorded markers. With a fixed number of bytes contained within each sector, you can store the number of bytes equal to the size of the sector times the number of sectors recorded on each track.

There is more than one track per disk. In fact there are two directions in which you can distribute the tracks: on different surfaces of the platter and across the surface. Typically, a disk has more than one surface. It usually has two surfaces and multiple platters. By placing a read/write head in close proximity of each surface, you can access this data by electrically switching heads and reading or writing data on this surface. This organization is known as a *cylinder* because you can visualize the track as physically extending through each platter forming a three dimensional cylinder of data. You can now store the number of bytes in each track times the number of tracks in each cylinder. By moving the heads, you can take advantage of the surface area by forming concentric rings or tracks on each surface and concentric cylinders across the platters. With this additional dimension of storage, you can store the number of bytes in each cylinder times the number of cylinders on each disk drive.

Floppy Disk Drives

To better understand how the logical MS-DOS file system organization relates to the physical disk, we need to select a specific disk type to study. Hard drives vary in architecture from vendor to vendor. Just look at the number of drive types in your computer's BIOS. Floppy disks are much more standard and easier to study. I'll discuss a 360Kb floppy disk.

Like the generic disk, the data organization on a floppy disk is broken into sectors, tracks, and cylinders. The platter on a floppy disk is made from mylar as opposed to aluminum or some other alloy. Because the mylar is very flexible when compared to the metal platter, the term "floppy" was applied to it. There is only a single platter with one or two surfaces, so a maximum of two tracks compose a cylinder. The number of cylinders vary, but most common is 40 or 80.

For a 360Kb floppy, the organization is nine sectors per track, two tracks per cylinder and forty cylinders per disk. Other architectures are possible, such as 720Kb and 1.44Mb floppy disks. These disks are also different physical sizes, typically 5 1/4-inch and 3 1/2-inch, but other sizes such as 8-inch and 3-inch existed in the past. With MS-DOS, the sector size is typically 512 bytes, but other sector sizes were also used in the past.

Logical Versus Physical

Both MS-DOS and DOS-C internally reference the disk drive as an array of fixed-size blocks of data. Each block of data in this array is sequentially numbered from zero to one less than the maximum size of the disk (in sectors). DOS-C and MS-DOS impose a structure upon this array of sectors in order to easily access data and organize files. This logical file system organization is shown in Figure 2.2. DOS-C and MS-DOS place a boot sector at the first sector on the disk because this is the best defined sector on the disk. This sector contains the code to load the operating system and a data structure that contains all the pertinent variables that define the disk geometry. For a 360Kb floppy disk, the next sector is a reserved sector and it's followed by two FAT areas, each two sectors long. The last area reserved by the file system is the root directory and it's 7 sectors long. The remainder of the disk is dedicated to data and subdirectories.

Mapping this logical structure into the physical disk requires some thought. In order to map the disk, you need to standardize how the logical sector numbers map into physical sector, cylinder, and track numbering. For MS-DOS and compatible operating systems, the first block is mapped into sector 1 of cylinder 0, track 1. The track number increases for every nine sectors on a 360Kb floppy, followed by

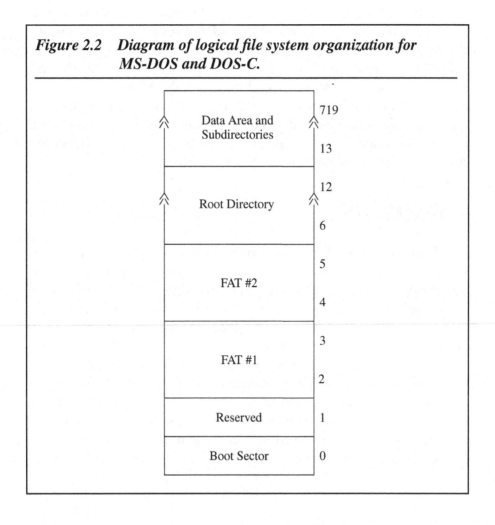

Figure 2.2   Diagram of logical file system organization for
MS-DOS and DOS-C.

increasing cylinder number. This mapping is illustrated in Figure 2.3. The file system structure does not take up a full cylinder and is spread out over two tracks.

Although not immediately apparent, this organization allows for the most rapid access when reading files sequentially. Each sequential sector can be accessed immediately upon reading the previous sector. When the last sector on the track (sector 9 for a 360Kb disk) is read, the first sector on the next track can be immediately accessed on the start of the next rotation. Once all sectors for every track in the cylinder are read (two tracks for a 360Kb disk), the next cylinder is read by physically moving the heads. This is important because there is a time penalty for stepping the disk heads on the order of several milliseconds. It's also important that the sectors follow each other so that you don't incur a penalty of a rotation of the disk before reading the next sector. This delay is also on the order of several milliseconds. Although the sectors are sequential on a floppy disk, some hardware is not fast enough, so the sectors are interlaced so that the sequential disk reads are completed in a minimum number of rotations. This is known as *sector interleave* and is necessary on some smaller XT-type systems.

MS-DOS API

As I discussed earlier, Microsoft based the early MS-DOS API on the CP/M model. This model enters all call parameters into one or more registers and makes a system call, possibly to a fixed address. MS-DOS offers two such system calls. One address is cs:05h. This system call is very nearly a duplicate of the CP/M system call. CP/M used the same address, and all of the initial MS-DOS v1.0 system calls (sometimes called the FCB system call, because of the reliance on File Control Blocks for disk I/O) are nearly identical to their CP/M counterparts. This model is primarily for the .com-type programs coded as small or tiny model programs. These program images also are similar to the CP/M-style executable image. Although the primary reason for this entry point is to accomodate .com programs, it is available to tiny- and small-model .exe programs as well.

Figure 2.3 **Diagram of logical file system organizaiton mapped into a physical disk.**

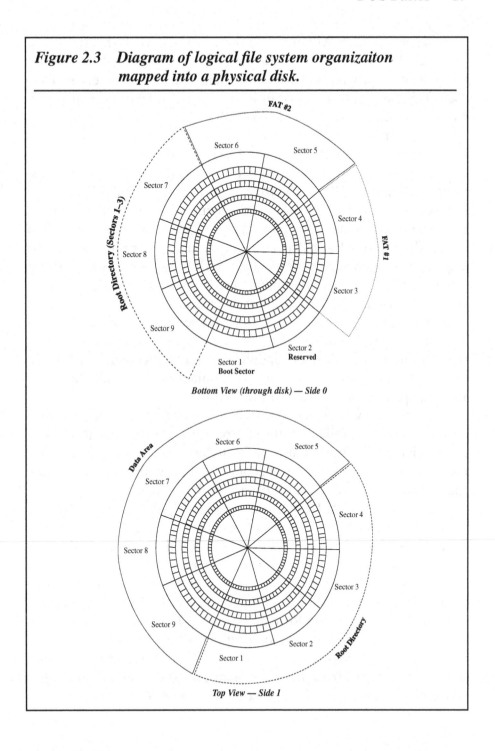

Another MS-DOS API entry similar to the CP/M style cs:05h is a call to 0:90h. As with the cs:05h call, this system call entry is for use by .exe-style programs that use medium and large models in their code. This entry point received little use, if any, and was actually incorrect in some earlier versions of MS-DOS.

By far the most used and widely accepted MS-DOS API is the int 2xh system calls. Unlike CP/M, the MS-DOS designers had software interrupts, such as int 21h, to work with. Using software interrupts causes the processor to push ip and cs register flags on its stack. It then looks up a new vector by translating xxh (by multiplying it by four) and placing the long-word vector found at the previously computed memory location into cs:ip. This is a very fast call and is highly flexible. Microsoft reserved the range of 20h through 3fh for use by MS-DOS and derivative operating systems. Other interrupts are similarly used by services outside MS-DOS, such as the system BIOS and application programs.

At the highest level, the Microsoft-reserved vectors of 20h through 31h allow a maximum of 18 software interrupts. However, by using values in registers to further subdivide each call, the technique allows for 13 API entry points with almost unlimited function calls for each. Table 2.1 presents a summary of these calls.

Through its development, MS-DOS was the recipient of three varieties of API call methods that use the software interrupt call technique. Each technique is simple to use and well documented. They do require, however, that the programmer become familiar with each and follow conventions. Otherwise, MS-DOS can break very easily.

The first type of API call is the simple software interrupt. System calls int 20h and int 24h–29h are examples of this type of call, which has only one defined function for each system call. Examples of the simplest calls are the terminate calls (Terminate Program and Terminate and Stay Resident). Through these system calls, programs return only simple status information to MS-DOS. Examples of slightly more complex calls are the Absolute Disk Read and Absolute Disk Write calls. As with the simpler calls, there is no modification to function behavior by the call, and the call performs only one function. These calls are only slightly more complex in that the system call returns to the caller after performing its function, allowing the application program to continue.

The second type of API call is characterized by multiple functions available through a single `int xxh` entry. The primary DOS API `int 21h` is an excellent example. To select a function, load register `ax` with a function number that corresponds to a unique system call. The application program loads other registers with variables applicable to the particular system call or pointers to an area of memory for system calls. MS-DOS then performs the operation and returns its results in registers or memory pointed to by registers.

Table 2.1   DOS `int` System Calls.

`int` **Number**	**Function**
20h	Terminate program
21h	System calls
22h	Termination address
23h	Break address
24h	Critical error
25h	Absolute disk read
26h	Absolute disk write
27h	Terminate and stay resident
28h	Idle handler
29h	Fast console I/O
2ah	Network handler
2bh	Not defined for DOS kernel
2ch	Not defined for DOS kernel
2dh	Not defined for DOS kernel
2eh	`command.com` reload
2fh	Multiplex interrupt
30h	`far` jump DOS entry
31h	Part of 31h

The MS-DOS `int 21h` system call is also the best documented example of this type of system call. Close examination of this API reveals that many functions contain not only a major function number but a minor function number as well. This allows greater expansion of system calls and also allows for logical grouping of related system calls. The IOCTL system call is the best example of this.

The third type of API call is the chained system call. Chained system calls used by MS-DOS extensions, such as installable file systems and network system calls, are similar to the techniques used to extend MS-DOS by third-party vendors. The multiplex interrupt, `int 2fh`, is an example of this call. When an application or MS-DOS uses a function installed into the multiplex chain, it loads a major and minor function number into the `ax` register. Each extension saves the call context and examines the `ax` register. If it matches, it performs the service and returns to the application or MS-DOS through an `iret` instruction. If it is not a function defined within the extension, MS-DOS restores the call context and passes control to the next extension's entry point stored during installation of the extension. Should no extension handle the call, MS-DOS performs a default action and returns to the application.

You need to study these API calls because a clear understanding of system call techniques will be very important later when you design your DOS-C API. Because the goal is to create a DOS-compatible operating system, you will need to respond to the same type of system calls in order to guarantee a level of compatibility. The exact mechanisms will be discussed in later chapters.

MS-DOS File System

The MS-DOS kernel supplies applications with a number of services. These services provide process management, memory management, file management, and miscellaneous services. Probably the greatest number of MS-DOS system calls relate or pertain to file management. This section explores the MS-DOS primary file system, the File Allocation Table (FAT) file system, and examines key details and data structures.

As I discussed earlier, the original MS-DOS purpose was to manage small, 160Kb floppy disks. Fixed disks with gigabyte storage capacities on a desktop computer and multimegabyte memory were unheard of. Instead, MS-DOS designers faced space restrictions both on disk and in memory. They also faced high rotational and head positioning latency times inherent to the 51/4-inch floppy disk drives. This forced MS-DOS designers to be creative in both disk layout and system call functionality. This creativity will be apparent as you closely examine the FAT file system.

Begin the examination of the FAT file system by examining a simple DOS floppy disk that contains only a root file system and is representative of a freshly formatted floppy disk. Its structure is flexible enough to be compatible with almost any removable media, regardless of size. Figure 2.4 illustrates a conceptual model of disk storage in an operating system compatible with MS-DOS. In MS-DOS, a logical disk is a sequential set of 512-byte sectors. The disk consists of a boot block,

Figure 2.4   Floppy disk storage model.

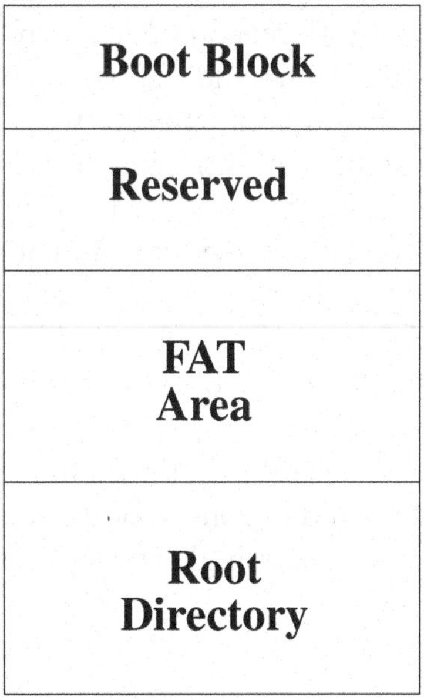

Boot Block
Reserved
FAT Area
Root Directory

followed by zero or more reserved sectors, and the file system. The boot and reserved sectors are for disk booting and disk maintenance, although the boot sector also carries file system information.

Following the boot-related area is one or more FAT areas. Each FAT consists of one or more sequential sectors in which each entry consists of a forward pointer in a linked list of pointers. Values of zero indicate free space, and other values indicate bad sectors, disk size, and disk type. The need for more than one FAT stems from the greatest weakness of the MS-DOS file system, the FAT itself. If a program crash destroys a FAT, no other information anywhere within the file system allows you to recover your files. Because disks are probably the component in a computer with the lowest reliability, redundant FATs improve the file system reliability compared to a single FAT.

Immediately following the FAT(s) is the root directory. In early MS-DOS days, the root directory was the only disk directory, similar to its CP/M cousin. All data relating to filename, location, creation date, file size, and file protection are contained within an entry in the directory. Starting with MS-DOS v2.0, the file system became hierarchical with other directories that extend from the root directory. However, there are differences between a root directory and other disk directories. The first significant difference is that the root directory does not contain an entry for "." (itself) and ".." (its parent) directories. This choice allowed for backward compatibility with systems running pre-2.0 versions. Another significant difference is that the root directory is fixed in length, although a subdirectory can be expanded. Again, compatibility with pre-2.0 versions forced this limitation.

Hard drives are very similar to the diskette in design. The biggest difference is that hard drives can support multiple disk images, which in turn can support more than one operating system. This concept borrows the idea of multiple volumes from other operating system designs. It improves on the concept by placing the multiple-volume concept outside the operating system and in control of the boot code. Each volume is called a partition and shares a single master boot record.

As with floppy disks, when an x86 system boots, it loads the first sector into memory. The difference between the floppy and hard disk is in the functionality of the boot code contained in this sector. The master boot record contains code that examines a partition table also contained in this sector. It determines the active partition and proceeds to load the first sector of this partition into memory. The remainder of the boot process is then exactly the same as for a floppy disk. By using this design, the file system remains identical for both floppy disks and hard disks.

However, certain on-disk changes to the data structure are necessary when certain physical sizes exceed certain structure member sizes. This occurs on drives when the physical size begins to exceed a 12-bit encoding limit in the FAT and a 16-bit limit on the device drivers. These limits challenged MS-DOS but its designers were smart enough to extend to 16-bit FAT entries and 32-bit block addressing to overcome these hurdles.

What you have seen so far is the principal MS-DOS file system architecture. Other file systems are supported by MS-DOS, such as the CD-ROM file system, but I will not cover them due to space limitations. However, I will further examine each of the file system components in detail so that you can design your DOS clone.

Boot Area

The first area of the disk, as seen earlier, is the boot area. This area consists of the boot sector plus zero or more reserved sectors. You already know that this sector executes first in order to start MS-DOS, and you know that every disk has one whether it is bootable or not. What you don't yet know is what the structure of this sector is and how MS-DOS knows how to access a given sector on different disk geometries.

Figure 2.5 shows a diagram of the boot sector.[†] Three areas are shown (although this varies for Master Boot sectors). Every disk boot must contain these three areas. The first area, called boot sector, contains an important disk structure with all information about the architecture of the disk. The second area is the body of the executable boot

---

[†] File system structures are presented as diagrams in this book to avoid confusion. The MS-DOS design is for 80x86 processors and ignores portability issues. As a result, all data structures contained on disk reflect 80x86 byte and word order. Additionally, all information published by Microsoft reflects the assembly language roots of MS-DOS through its presentation as assembly language data structures. Later chapters address portability issues and techniques. All data read from and written to a storage medium respects MS-DOS conventions, but all in-memory data structures are native C structures for efficiency.

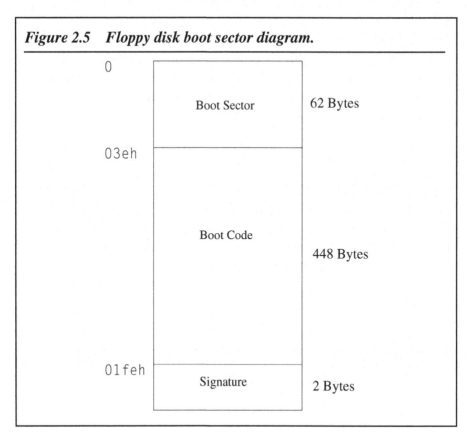

Figure 2.5   Floppy disk boot sector diagram.

code. The final area is a signature word, 0x55aa, used to identify this sector as a valid boot sector.

The earlier discussion centered on how MS-DOS boots. In it, I discussed loading the boot sector into location 0:7c00h and transferring control to this address. I also spoke about various disk geometries.

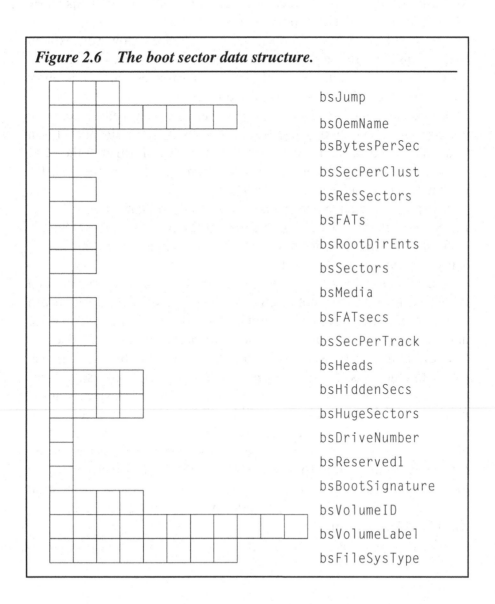

Figure 2.6 The boot sector data structure.

bsJump

bsOemName

bsBytesPerSec

bsSecPerClust

bsResSectors

bsFATs

bsRootDirEnts

bsSectors

bsMedia

bsFATsecs

bsSecPerTrack

bsHeads

bsHiddenSecs

bsHugeSectors

bsDriveNumber

bsReserved1

bsBootSignature

bsVolumeID

bsVolumeLabel

bsFileSysType

I will use the boot sector data structure, shown in Figure 2.6, to explain how this is possible.

The first entry in this table is an area called bsJump, which is a reserved entry that contains an Intel 3-byte jump: either e9 xx xx absolute jump or eb xx 90 relative jump. Fortunately, the ugliness of the Microsoft design for Intel processors rears its head only with this entry. The sequences are actually Intel opcode for an absolute jump or a relative jump followed by a no-op instruction. It does, however, explain how the disk information (in the boot sector area) placed ahead of the boot code does not interfere with the booting process since this entry points to the actual boot code entry address.

The remainder of the data structure is much of what you expect from the boot area. The next entry, bsOemName, is 8 bytes allocated to an OEM name and version. This entry identifies who formatted the disk, not which operating system is currently running. It is on the disk for information only and serves no other purpose.

Starting with the next entry, the boot sector duplicates the MS-DOS BIOS Parameter Block (BPB). The first BPB entry, bsBytesPerSec, is a 16-bit entry that specifies how many bytes are in a sector. This is typically 512, but can be different.

The next area, bsSecPerClust, indicates how many sectors are in a cluster. Although I have not discussed the cluster yet, this is where you first encounter it. A *cluster* is a group of sectors used as a fundamental unit of storage allocation and serves two purposes. First, it places a level of indirection between the operating system and the physical disk media format. This allows changes in disk and operating system technologies to be mutually exclusive. Second, it allows disk sizes to grow much larger than would be possible with just a sector allocation scheme. An example is a typical hard disk allocation scheme. These disks typically use 4096-byte clusters, allocating 8 sectors per cluster. If the addressing limit is 16 bits, the limit is 33Mb of storage using sector allocation but increases to 268Mb using an 8Kb cluster allocation scheme. With 32 bits, sector allocation yields approximately 2200Gb and 17,600Gb respectively.

The entry `bsResSector` specifies how many reserved sectors are on the drive or disk, including the boot sector. This entry is useful for both the operating system and the boot program. As you will see later, the operating system uses this entry in computing the location of key disk data structures. Boot can use this entry to compute how many additional sectors to load into memory for more complex booting schemes. An example of such a scheme may be booting multiple operating systems. It is possible for boot to load a secondary, more complex boot program that would then proceed to present a menu and load the selected operating system. Other schemes are also possible, limited only by the developer's imagination.

Jumping ahead, `bsMedia` is the media descriptor. This scheme was an early attempt at automated identification of the media type that is present in a drive. MS-DOS passes this byte, by specification, to the device driver. Originally, this entry switched a drive from low to high density. It is, however, an incomplete scheme where the same byte appears to specify more than one type of media to the same device driver. Although Microsoft may use it, I will only pass it to the device driver for compatibility purposes.

The entries `bsFATs`, `bsRootDirEnts`, `bsFATsecs`, `bsSecPerTrack`, and `bsHeads`, when combined with `bsResSector`, are used to translate a logical sector number to a physical sector location. For example, to get to the primary FAT disk logical sector, start at logical sector `bsResSector` (logical sectors start at 0). To get to the secondary FAT logical sector, start at `bsResSector + bsFATsecs`. An interesting note is that more than two FATs are permitted on any disk, implied by `bsFATs`. So, if you want to get to the nth FAT, start at

$$bsResSector + (n*bsFATsecs)$$

resulting in the generalized FAT algorithm.

From Figure 2.4, you saw that the root directory followed the FAT areas. You can compute the start of the root directory with the algorithm

$$\text{bsResSector} + (\text{bsFATs} * \text{bsFATsecs}).$$

Finally, any data sector you need to find starts at the logical sector

$$\text{log } sector = \text{bsResSector} + (\text{bsFATs} * \text{bsFATsecs})$$
$$+ (32 * \text{bsRootDirEnts} / \text{bsBytesPerSec}).$$

Because I have been talking about logical sectors, you will later see that this is the method MS-DOS uses to convey a desired sector to the device driver. The de facto DOS convention for physical translation is the sequential cylinder model, where each cylinder is composed of a number of tracks, and cylinders increase with logical sector numbering. To translate a logical sector to a physical cylinder, track, and sector model, the cylinder number is computed as the integral computation

$$\text{log } sector / (\text{bsSecPerTrack} * \text{bsHeads}).$$

The track number is computed by

$$\text{log } sector \text{ mod } (\text{bsSecPerTrack} * \text{bsHeads}) / \text{bsSecPerTrack}.$$

Finally, the physical sector number is

$$((\text{log } sector \text{ mod } (\text{bsSecPerTrack} * \text{bsHeads}))$$
$$* \text{mod } \text{bsSecPerTrack}) + 1.$$

(Physical sector numbering starts at 1.)

There are only two other key entries, bsSectors and bsHugeSectors. These entries are used for error checking by the device driver. If a logical sector number is greater than or equal to bsSectors or bsHugeSectors, then an error condition exists. The reason for two separate entries is historical. Originally, bsSectors was the only entry governing disk size.

It is limited to 16 bits, which is insufficient for larger hard disk drives. During its evolution, MS-DOS gained a second entry to handle these larger drives, bsHugeSectors. The method used to decide which entry to use is a simple algorithm: look at bsSectors, if it is zero, use bsHugeSectors, else use bsSectors.

The remaining entries, bsBootSignature, bsVolumeLabel, bsReserved1, bsVolumeID, bsDriveNumber, and bsFileSysType, deal with disk volume identification or are internal data structure members. The remaining sector, bsBootSignature, is used in other calls, and all of the remaining data structure members are informational only.

bsHiddenSecs is an additional entry that allows MS-DOS to hide, or reserve, additional sections that are hidden from the file system. However, bsHiddenSecs is rarely used.

FAT Area

Interestingly enough, the FAT is probably the best-documented area in all of MS-DOS. Frankly, I don't know why, but I will review it here in order to present a complete MS-DOS file system discussion. For those of you who find this boring, skip ahead.

All files contained on the basic DOS file system are composed of three distinct parts: a directory entry, a linked list in a FAT, and a collection of disk sectors that contain the information in the file. I will discuss the directory entry later, but for the purposes in this section, the directory entry contains the means by which MS-DOS identifies the file by name and by a number used as an index into the FAT array. For example, when MS-DOS opens a file for read access, it finds the file by matching the filename and retrieves the starting FAT index. In essence, this is the only action necessary to open a file, although MS-DOS maintains other internal data structures that are also updated.

The FAT and the data area have a one-to-one relationship. This relationship allows MS-DOS to keep track of the data clusters in an efficient manner. By definition, each FAT entry corresponds to its equivalent sequential cluster. For example, any reference to cluster number 3 also refers to FAT entry number 3. The FAT functions as a linked list. If you want to find the cluster that follows cluster number 3, you look at the FAT entry number 3 and get its content. This number now gives the next data cluster number. Figure 2.7 illustrates how this works.

In the example, you see two files. The first file, autoexec.bat, is probably very close to one you may have on your boot disk. When MS-DOS opens the file, it first goes to the directory and begins a linear search for the filename autoexec.bat by looking for a string match. When MS-DOS finds a string match, it performs some internal updates to the file table and picks up the entry for the starting disk cluster, in this case 2. An interesting note about this entry, all clusters on a disk begin with cluster number 2, with 0 reserved as a free cluster indicator and 1 reserved for MS-DOS. This is important when you begin your design.

If you now instruct MS-DOS to read the file sequentially, you will read all sectors in cluster number 2 first. MS-DOS does this in one of three ways, depending on the number of bytes requested and the current

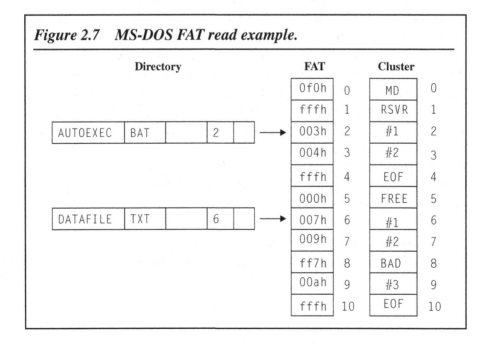

Figure 2.7   MS-DOS FAT read example.

position of the MS-DOS internal file pointer. In the first case, when you read less than a whole sector and the file pointer is at an arbitrary position within the file, MS-DOS reads a sector from the cluster into a buffer and transfers the requested number of bytes to memory. In the second case, where the file pointer is at the start of a sector boundary and the requested number of bytes is a multiple of a sector, MS-DOS transfers the data directly into memory. In the third case, the internal pointer is at an arbitrary position and the requested data transfer is greater than a sector. In this case, MS-DOS performs a combination of actions of the first two cases and continues until a new cluster is needed.

When MS-DOS begins to read the next cluster, it needs to find its physical location on disk. It does this by looking at the FAT in the entry corresponding to the current cluster number. The number it reads corresponds to the next entry to be read. In this fashion, each FAT entry corresponding to a physical cluster forms a forward link for the corresponding file data area. This process continues until MS-DOS encounters an end-of-file marker. The example in Figure 2.7 shows this for the third cluster of autoexec.bat. The sequence of clusters is 2, 3, and 4. FAT entry 4 contains an fffh, which signals MS-DOS that this is the last cluster of the file (by definition, any entry in the range of ff8h through fffh indicates the end-of-file). Table 2.2 lists all possible values contained in a FAT entry and their meanings. One note — this

Table 2.2   FAT entry values and meanings.

Value	Meaning
(0)000h	Free cluster
(0)001h	Not used
(0)002h-(f)fefh	Data cluster indices
(f)ff0h-(f)ff6h	Reserved
(f)ff7h	Bad Cluster, used for bad sector mapping
(f)ff8h-(f)fffh	Last cluster

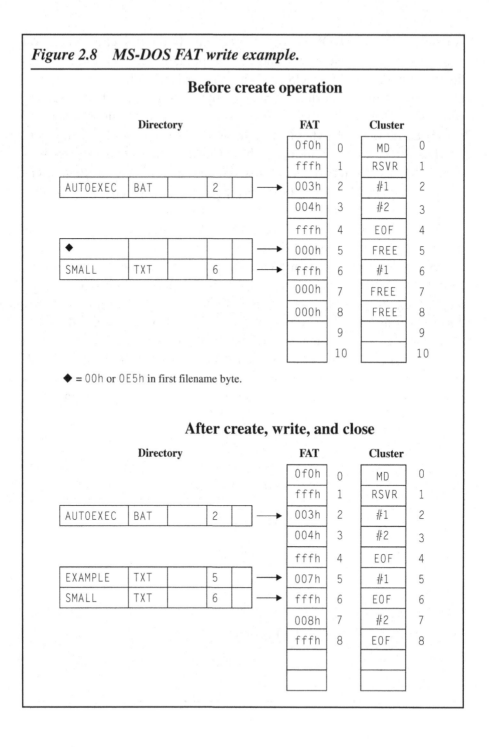

Figure 2.8   MS-DOS FAT write example.

Before create operation

				FAT		**Cluster**

◆ = 00h or 0E5h in first filename byte.

After create, write, and close

discussion uses 12-bit values, which correspond to the type of FAT typically found on floppy disks and small hard disks. You quickly run out of range on larger hard disks, so a 16-bit FAT is used for these. This discussion holds true for 16-bit FATs, just sign extend the numbers.

For a write, the operation is somewhat more complex because file position plays a big part in how write works. However, you can examine some fundamental MS-DOS FAT principals for write with a simple sequential file create and write example. Figure 2.8 illustrates this example.

In the example, you start by creating example.txt, the example file. First, MS-DOS searches for a free directory entry to begin its allocation. It does this by searching for either a 00h or 0e5h for the first character of the entry (I will discuss their meaning in the next section). When it's found, internal file data structures are updated and the operation is complete. You can now begin the write operation, and this is when the FAT fun starts. As soon as you do a write, the very first cluster allocation occurs. In the example, cluster 5 is the first available cluster and is allocated to our example file. When you have written exactly one cluster size plus one byte, the next cluster is allocated. In the example, cluster 6 is occupied by a cluster belonging to small.txt, so the search returns cluster 7. The FAT forward link entry for cluster 5 now contains 7 and the writing continues until you finally close the file. In the example, this occurs somewhere in the third cluster. MS-DOS updates its data structures and places an EOF indicator in the FAT forward link entry, relative cluster 3 or physical cluster 8.

I have been very careful in discussing the write operation by avoiding all references to time in my discussion. As MS-DOS matured, it seems that the timing of the actual physical write to disk changed. Also, I wanted to keep the discussion very generic because in the later design you will want to add some robustness to DOS-C, and you will see that the method used to update the physical disk dictates exactly how much of a file is recoverable in the event of a hardware or power failure during the write operation.

Directory Area and Directories

The last part of the FAT file system discussion is the directory area. I have already covered both the data area cluster mapping and the FAT forward link and shown how an MS-DOS file is composed of three components. The directory is the glue that holds the file together. In MS-DOS, there are two types of directories, the root directory and subdirectories. I draw this distinction because there are some subtle differences between them.

The first distinction is that every disk must have a root directory. The root directory is the starting point for all directories and files. Whenever a search down any path is started, starting from the root directory is guaranteed to get you to the exact file or directory.

Another difference is in the contents of the root directory. A root directory contains entries not only for files and subdirectories but also for the volume label that appears at the top of a dir command output. It is also why the volume label has a limit of 11 characters: It must fit into the space normally reserved for filename (8 bytes) and extension (3 bytes). Additionally, it also does not contain an entry for "." (self) and ".." (parent), whereas all subdirectories must always contain these entries.

The final distinctions between the root directory and subdirectories are location and size. The root directory consists of a sequential set of sectors starting at a fixed location on the disk, whereas a subdirectory is actually a special MS-DOS file that can only be accessed by MS-DOS itself for both read and write. The size of the root directory is fixed by an entry in the boot area, and subdirectories can grow in much the same way a file can. As you will see later, these differences are critical in the design of directory handlers within DOS-C.

Having noted these critical differences, I can now state what is common in all types of directories. Any MS-DOS FAT file system directory is stored as a linear array of directory entries. The rules for use of a directory entry are common for all file types. Figure 2.9 illustrates a directory entry structure.

There are not very many directory entry fields, reflecting the simple function that a directory entry performs within MS-DOS. The directory entry performs four functions: name the file, specify access rights, hold time stamps, and specify the file location and size.

When it comes to filenames, MS-DOS borrows from its CP/M heritage by borrowing the filename and file extension method. MS-DOS limits the filename to an 8-byte entry and the extension to a 3-byte entry. Each entry is left justified and space filled. Hence the file "abc" is stored as "ABC⎵⎵⎵⎵⎵", the file "B.C" is "B⎵⎵⎵⎵⎵⎵C⎵⎵", and the file "FileName.Txt" is "FILENAMETXT". There are no typographical errors in the previous example; all filenames and extensions are converted to upper case by MS-DOS, further simplifying filename match algorithms. One final note, an unused field is denoted by a 00h and a deleted file by a 0e5h in the first location of deName. This distinction is important, because MS-DOS stops its linear search in a directory when it encounters 00h. It is an interesting way of making files disappear on an MS-DOS disk.

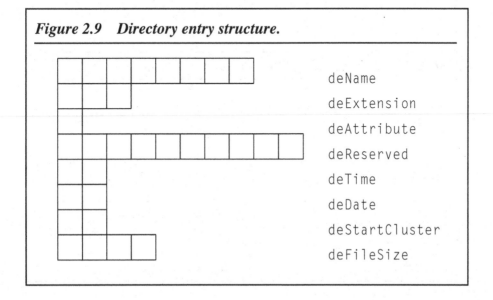

Figure 2.9   Directory entry structure.

The deAttribute entry restricts file access. In this field, a set bit indicates that an attribute is active. Table 2.3 lists these attributes. One interesting note — the ATTR_ARCHIVE bit is actually a "modified" bit set, whenever MS-DOS does a write operation to the file. In this way, MS-DOS knows that a file with this bit set needs to be archived.

For people who are familiar with UNIX and other operating systems, these attribute bits and file restrictions seem primitive. In essence, they are sufficient for a small, personal system. I will revisit these bits and their meanings later in the design of DOS-C.

Time stamping is another function of the directory entry for which MS-DOS uses the last file access that modified the file to generate the date and time fields. The choice of encoding method was the preference of the operating system architect. Unlike UNIX where the time stamp is seconds from an epoch, the MS-DOS time stamp consists of both date and time. Like UNIX, the MS-DOS time stamp is 32 bits wide, broken into a 16-bit date field and a 16-bit time field. These two fields are further broken down into bit fields composed of the logical parts of the element, such as hours:minutes:seconds and year:month:day, as shown in Figure 2.10. Time stamps will become a critical data item to update when I visit functions such as create, write, and close later in our design.

Table 2.3   Attribute bits of the deAttribute entry.

Mnemonic	Value	Meaning
ATTR_READONLY	01h	A read-only file
ATTR_HIDDEN	02h	Hidden or directory (can't delete)
ATTR_SYSTEM	04h	System or directory (can't delete)
ATTR_VOLUME	08h	Volume label, not a file
ATTR_DIRECTORY	10h	A directory
ATTR_ARCHIVE	20h	Modified file

The final ties, and probably the most important, are the file location links. These fields link the FAT and data clusters to the directory entry. The field deStartCluster is the FAT link to the first FAT entry. This is in essence the anchor for the file. The entry deFileSize completes the file description by specifying the size of the file in bytes. With this entry, the file becomes independent of file system architecture or cluster size. It does, however, limit physical file sizes to 4.3Gb.

The design of the directory is pivotal in any program that manipulates a FAT disk, and it is no exception in DOS-C. Design tradeoffs concerning the timing of directory updates will affect file system robustness. There are also design issues regarding when to switch between 16-bit and 32-bit FAT tables, etc. I will examine these in later chapters and use the information presented here during the design.

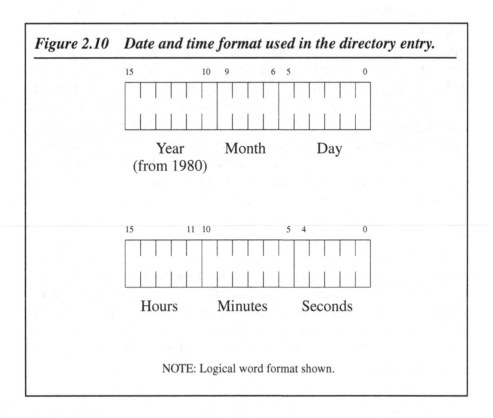

Figure 2.10   Date and time format used in the directory entry.

NOTE: Logical word format shown.

MS-DOS Memory Management

In all except the simplest operating systems, there exists one form or another of memory management. The reason is simple: memory is yet another computer resource given to the operating system to maintain. How an operating system performs this management varies drastically among systems. For example, large operating systems, such as UNIX, provide memory management in multiple forms. One is for program use as part of a heap management scheme. Another is for the program itself, either maintaining it for program loading or dispensing it in a fashion that gives the program the illusion of much more memory than is physically available. MS-DOS manages memory using a much simpler scheme but still manages program heap requirements and program loading space.

I must admit that the developers of MS-DOS did the best they could with a poor architecture. In many computers other than those based on the original PC and XT architecture, there is some form of a hardware memory management unit, which provides two functions: protection and relocation. On these machines, protection is achieved by disallowing writes to occur based on some table, either in memory or within the memory management unit itself. Typically, the write protection can be specified over portions of memory. This capability helps make these other machines more resilient to occurrences of wild pointers, pathological programs, etc., protecting other programs and the operating system in the process. It is much harder for application programs to crash these machines. Relocation allows the operating system to move physical memory around and make it look as though it is contiguous logical memory. Memory becomes available to a program in either a straightforward manner for allocation or on demand for virtual memory.

At the time of the development of the PC, the desktop computer market was based on simple systems running with simple microprocessors. When IBM decided to enter the market, it decided to emulate the simpler computers rather than introduce proven technology that it had used on its mainframes and minis. This unfortunate choice has always been a weak spot for MS-DOS, because it suffered attacks from critics for being vulnerable to crashes.

The developers of MS-DOS did, however, introduce some forms of memory management through the use of a memory block linked list that consists of a control header, known as a memory control block or arenaheader, and the allocated (or free) memory. (By now, it should be obvious that the linked list is the data structure of choice for MS-DOS.) To their credit, they did the best with what they had, and the MS-DOS memory management scheme does provide a fair degree of control.

Figure 2.11 illustrates the MS-DOS arenaheader. Throughout the history of MS-DOS, the arenaheader has been referred to as the MCB or Memory Control Block (they are the same data structure). This data structure is in front of each allocated and free memory block. In the first 640Kb, these control blocks contain an "M" in arenaSignature for all headers except the last one, which contains a "Z". The next field, arenaOwner, contains the segment value of the PSP (Program Segment Prefix) when it belongs to a user or contains a 0008h if it belongs to MS-DOS. This value identifies the block's owner and is used to dispose of memory when a program terminates. The next entry, arenaSize, is used to indicate the size in paragraphs of the memory block. arenaReserved is currently unused within MS-DOS and is reserved for potential future use. The final area, arenaName, contains either the name of the program that it belongs to, or a special name such as "UMB" to indicate a special arena. When MS-DOS divides a special arena into subarenas, it applies the same arenaheader and management techniques to these areas.

Figure 2.11   The arenaheader data structure.

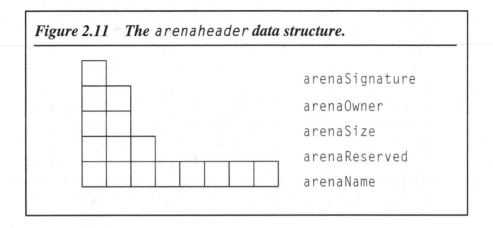

arenaSignature

arenaOwner

arenaSize

arenaReserved

arenaName

MS-DOS also provides a number of API calls to manage the arenas. The basic management provided is simple allocate and dispose. Calls are also used to set allocation strategy, such as first fit, last fit, and best fit. Additionally, there are calls that detect and manipulate the upper memory blocks.

MS-DOS uses special arena calls to further subdivide an arena when used for special MS-DOS applications. One special arena usage is for special storage allocations specified in config.sys, such as new stacks, File Control Blocks (FCBs), and device drivers. Another special arena usage is to identify and subdivide the upper memory blocks, which MS-DOS uses to load device drivers and the kernel itself into memory above the conventional 640Kb limit.

MS-DOS Task Management

Task management in MS-DOS is very simple. There are only three methods of loading a program: load and execute, load, and load overlay. Readers who are familiar with other operating systems may find the limited selection a surprise, but they represent the minimum set necessary for a small operating system. When combined with the Terminate Program and Keep Program system calls, the load methods offer a powerful set of system calls.

I want to point out to the reader that although MS-DOS offers a small selection of system calls that handle program execution, there is a great deal of work that MS-DOS must perform for these calls. Examine the Load and Execute Program system call as an example. The first thing that MS-DOS does is look for the executable file and identify what type of executable file it is: .com or .exe. It does this by examining the first two bytes. If it sees the correct .exe signature, it proceeds to load the file and do the correct relocation by correcting all segment references contained within the file. Otherwise, it assumes that it's a .com file and just loads the file directly into memory. It then creates the PSP and populates it with all inherited file handles, updates the termination address, Ctrl-C handler, and critical error handler. Next, the starting address is determined, a program stack is allocated, and machine registers are loaded. At this point, control is transferred to the loaded program. Much of the same effort goes into the Load system call. The only difference between the two is that transfer of control does not take place for Load. Another method is Load Overlay, but this only loads a program without changing the PSP or special handlers.

When program execution completes, two system calls are available: Terminate Program and Keep Program. There is considerable difference between these two calls. Both system calls store a program return value available to the parent program, but they soon differ in function. The Terminate system call proceeds to undo everything done by the Load system calls. It must begin by replacing the termination address, Ctrl-C handler, and critical error handlers with those of the parent and go through the PSP and operate on the child's file handles. After all this is complete, it returns the program's memory back to the memory pool, effectively destroying the program image. Control is then transferred back to the parent program and execution continues within the parent. In contrast, the Keep Program system call must omit destroying the image, closing files, etc. The reason for this is simple: Whereas other operating systems have the concept of daemons built into them — for example, UNIX uses daemons for all of its networking, `cron` program scheduling, line printer functions, and other tasks — the only way to accomplish a similar task in MS-DOS is to terminate the program and keep its image in memory. The technique is known as TSR or Terminate and Stay Resident. It is for this reason that Keep Program must keep the image and associated data structures, such as the PSP, intact, since the program may be triggered by an external event. With this in mind, the differences become clear. Keep Program only stores the return code and returns control to the parent program. This guarantees that the program image remains intact.

MS-DOS Device Drivers

The MS-DOS device drivers are simple in design but difficult to implement in C, not because the code is complex, but because the data structure used to communicate with the device driver is actually designed for assembly language and results in a very complex union to implement. The major contributing factor is that the structure implements a common beginning for control information, but the following data areas vary for each call. I will examine the mechanism used to control driver entry points and a portion of the common data structure.

The design of MS-DOS device drivers is very much tied to the x86 architecture. If you examine Figure 2.12, you will see two sets of pointers that are different in size. The first pointer, dhLink, links together all the device drivers used by MS-DOS, both internal and loaded through config.sys. The other two pointers, dhStrategy and dhInterrupt, are offsets into the two driver entry points. For these calls, the device driver segment is assumed to be the segment for the call, restricting the device driver size to one segment.

The two names of the entry points, dhStrategy and dhInterrupt, suggest device drivers similar to those used in multitasking operating systems. Unfortunately, when you examine the way the typical MS-DOS device driver is written, those names are very misleading. In other operating systems, the dhStrategy entry would queue up a request and

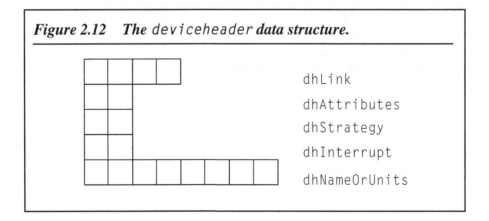

Figure 2.12   The device header **data structure.**

dhLink

dhAttributes

dhStrategy

dhInterrupt

dhNameOrUnits

possibly start the transfer, and dhInterrupt would be the interrupt handler that might terminate the data transfer. MS-DOS designers never did get to complete this. In MS-DOS, dhStrategy is called first. When it is called, the device driver usually just stores the request pointer. The driver entry, dhInterrupt, is then called immediately afterwards where the actual driver call is implemented.

The requestheader is the common data structure that prefixes all device driver requests (Figure 2.13). When the device driver is called, a far pointer to the request is passed to the device driver. Appended to the request area is a variable data structure that is unique for each system call. Because of this variability, rhLength indicates the length, in bytes, of the combined requestheader and unique data structure. To allow the device driver to handle more than one device, rhUnit specifies the device number that the driver is to operate on. The entry rhFunction specifies the function, such as read or write, requested of the driver. Finally, when the operation is complete, rhStatus returns the completion status. The entry rhReserved is unused at this time and is reserved for future use.

Of the remaining deviceheader entries (Figure 2.12), the dhAttributes and dhNameOrUnits entries are used for informational purposes. dhAttributes is used to convey information to MS-DOS about the device driver. It is a bit field that indicates whether the device driver is for a block or character device, whether it is stdin or stdout, etc. The final field, dhNameOrUnits, contains either the name of the device if it is a character device or the number of units handled by this device for a block device.

Figure 2.13   The requestheader **data structure.**

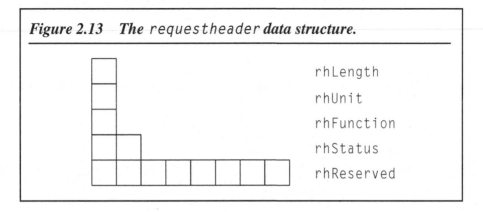

rhLength

rhUnit

rhFunction

rhStatus

rhReserved

Unfortunately, you need to learn many more details before attempting to write a device driver. This task is beyond the scope of this book, but the topic is well documented by Microsoft and other publishers. I highly recommend that you refer to these sources before writing your own device drivers. I will go into the details of each device driver later in this book and cover what calls are handled and how.

What I Am Not Covering

As you can see by now, even a small desktop operating system like MS-DOS can be complex. You have seen that the API has different variations, a file has two parts hidden by the operating system, etc. However, there are parts of MS-DOS that I am not covering because I am not implementing them in DOS-C. I am not covering installable file systems because of DOS-C design limitations. Also, I will be taking liberty with internal data structures and error handling. With this information in hand, I will start the DOS-C design and implementation.

Chapter 3

Bird's-Eye View
of DOS-C

Previous chapters examined the architecture of MS-DOS and looked at some details of the operating system. They also created a model of MS-DOS and defined the minimum set of programs that form MS-DOS. This chapter will take a look at the overall architecture and the major components of DOS-C and will show its relationship to the MS-DOS model. This chapter will also examine the dynamics of booting DOS-C and explain the design behind the techniques used.

One major difference that you will see throughout this chapter is that the DOS-C design does not worry about smaller systems. This is really a reflection of the times. DOS-C expects to work in systems with at least 256Kb of memory, whereas MS-DOS expected only 64Kb (although the current kernel is much larger than the minimum memory requirements of MS-DOS v1.0). Also, you no longer need to worry about saving disk space for small floppy disks because you can expect systems that have over seven times more storage than the original PC disks. Finally, I want to use a high-level language in the implementation so I won't waste time trying various assembly language techniques to minimize size or maximize speed. As you will see in the final kernel, the penalty you pay for the convenience of rapid coding in a higher level language is far less than some programmers would like you to believe. So with this in mind, start examining the architecture of DOS-C.

Basic DOS-C Architecture

Chapter 2 outlined an experiment to determine the minimum set of programs necessary to run MS-DOS to limit the scope of the project, since individual programmers have to learn by example without the resources of a megacorporation with 200 programmers available to tackle the job. The result: MS-DOS needs, as a minimum, boot, io.sys, msdos.sys, and command.com to operate. These four components are the necessary set for MS-DOS, but this chapter will re-examine each component's functionality and broaden the definition to improve upon the design.

The boot sector is the first sector placed onto the diskette. You already know that all PC-style computers load and execute the boot sector in order to boot any operating system. As a result, you don't have much latitude in the design for boot. Also, with less than 512 bytes to use for the boot code, you won't be performing many miracles. We do have a fair amount of freedom in the next stage.

As you may recall, boot loads io.sys, which proceeds to initialize its internal device drivers. io.sys continues to initialize its device drivers by loading msdos.sys and instructing it to initialize itself. io.sys then uses msdos.sys to read config.sys, if present. When io.sys completes the processing of config.sys, it loads command.com and finally transfers control to it, allowing autoexec.bat to execute and ending up at the command line prompt. This is a good description of what MS-DOS uses, but you need to broaden the definition somewhat so that you can get a better picture of exactly what is happening.

If you step back and re-examine the operations, you can make some generalizations that will help define what you will do. You saw that boot loads a program that loads the full kernel. This loaded program is aware of a limited set of file system functions and knows how to read executable files into memory. If you take this generalization and look back into some computer science texts, you will find this to be the description of an Intermediate Program Loader or IPL. You will use this description of stage two of the booting process. You can now state that for DOS-C, the second stage is an IPL whose function is to load the final kernel.

You also see from the previous description that when stage two is complete, the complete kernel has been loaded into memory and it has initialized its internal data structures. Again, if you generalize this description, you can say that the IPL loads the kernel and transfers control to it. This is now stage three of the boot process.

The last part of the previous example then shows that command.com is loaded and control is transferred to it. Once again you will step back and generalize this stage. What you can say is that the kernel, after initialization, loads a user-level program that eventually brings you to the user interface.

These four stages of booting also give nice logical breaks for defining the four fundamental components. As stated earlier, the first component is boot, since its behavior is largely dictated by the computer architecture and not the operating system itself. The second component is an IPL that is aware of the file system and the structure of the kernel and loads the kernel into memory. The third component is the kernel, responsible for all the system calls and resource management of the operating system. The final component is the command line interpreter.

Now take these four components and make some decisions about the nature of each. You know that the boot must be in the first sector of the disk and that it must contain certain data structures. It is a binary image and will utilize the BIOS for communications. It will look for the IPL and load it into memory.

The IPL will be aware of the file system and be capable of loading an executable image into memory. If you look at this simple description, you see that you can say that the IPL is a specialized version of the operating system. This is an excellent way to design it, since you can concentrate your efforts on the kernel itself and use the C compiler's preprocessor to conditionally exclude unnecessary portions of the kernel. Because you know that boot is limited in capability, you will make the IPL a binary image easily loaded into memory and call it `ipl.sys`. You can also take advantage of its derivation from the kernel and give it the capability to load `.exe` files, yielding tremendous flexibility in the requirements for the kernel. You can load one or more files into memory, relocate them dynamically during the load process, and save development time since you will need to copy only one or more `.exe` files to the boot diskette in order to change the kernel.

For DOS-C, you will limit the kernel to a single file, `kernel.exe`, so that you will not need to get fancy with the development process. It will be a single executable, with full relocation capability, and you can generate it using standard DOS compilers and linkers. It will initialize itself and load the user-level programs to start the user interface.

Because you are modeling MS-DOS, the final component will borrow from MS-DOS and will be `command.com`. You do not necessarily need to do this, as you could have loaded an `.exe` version of `command` or an `init` similar to UNIX that brings the operating system to a multi-user state. However, so many programs have been written with the concept that the command line interpreter is loaded next and that it is called `command.com` that changing this step may only complicate things. So you will follow closely MS-DOS and end with `command.com` executing `autoexec.bat` to give the user interface.

Booting DOS-C

You have generalized the MS-DOS boot stages and the basic partitioning of the necessary operating system functions into DOS-C files. Examine the dynamics of booting DOS-C. Figure 3.1 shows the boot stages for DOS-C. Four distinct stages are involved in the boot, corresponding to the four minimum components necessary for DOS-C. This approach simplifies the design through modularization. By confining each stage to its own unique module, interaction is confined to the interface layers. A bug that may exist in `kernel.exe` does not require changes in `ipl.sys` nor will any change in size of either `ipl.sys` or `kernel.exe` affect each other.

The computer's ROM BIOS initiates the first stage. This stage is identical to the way almost any PC-compatible operating system starts out. The BIOS fixes the location of boot in memory and the initial `cs:ip`. For MS-DOS, IBM placed the location of boot at `0000:7c00h`

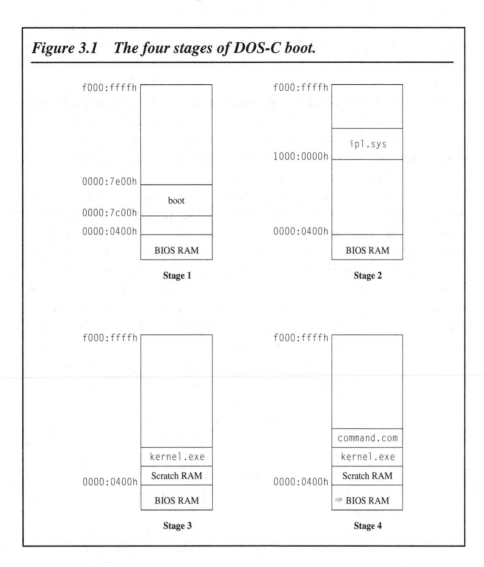

Figure 3.1   *The four stages of DOS-C boot.*

to allow 30Kb for loading `io.sys` and `msdos.sys`. DOS-C does not need boot to load in this location because all of the following stages are either position independent or relocatable, and you can judiciously juggle the following stages for any size kernel.

As a neutral choice, load the next stage, `ipl.sys`, into memory at `1000:0000h`. There is no magic to this address: it can be almost anywhere. You can dynamically locate `ipl.sys` anywhere in memory because you have built it from the kernel source code using the x86 small model. This limits the total size of `ipl.sys` to only 64Kb. By limiting the scope of the IPL, you can easily fit its functionality into this memory size.

`ipl.sys` loads `kernel.exe` and is aware of MS-DOS file systems. However, its lifetime is short and its purpose is simple: load the kernel. In order to reduce development time, it is logical to derive the IPL from the kernel source by reducing the kernel functionality to only what is needed to load the kernel. You will see how I have derived `ipl.sys` from the kernel source code when you examine the source code, but for now I will outline what you don't need. For one thing, you won't need any sort of API since you won't have any other program using the internal functions. You can also reduce the size of `ipl.sys` by eliminating functions you don't need. For example, because it is strictly a loader there is no need for functions that write or delete. You only need functions that open, read, and close. You can further simplify `ipl.sys` by restricting it to only one loader. You can always switch to the other or include both if you decide to do so in the future. Realistically, if you choose the `.exe` loader you are not restricted to a programming model, and therefore this is the choice for the design. Also, you don't need all of the device drivers. It does not make sense to have a line printer or serial port driver if the primary function is booting.

You also should recognize the potential behind an IPL boot process. You can design a debugger or menu system into the IPL to choose the program you intend to load. Either feature enhances the debugging process, since you can stop after loading the kernel and start debugging, or you can decide to change the name of the experimental kernel and select it at runtime.

Stage three starts once the kernel is loaded into memory. The kernel starts at `main()`, just as any C program will, and begins by initializing itself. This includes locating all of memory, building arenas, and loading the proper addresses into the interrupt vectors. After initialization, processing `config.sys` can occur, but you won't do that for this version. Finally, the kernel searches for and loads `command.com` into memory and makes this the granddaddy of all programs. If `command.com` returns, the kernel will shut down and you will need to reboot. This is no different than the case where MS-DOS cannot find `command.com` and issues the error message "Bad or missing command interpreter," except that the approach is cleaner.

When `command.com` is successfully loaded, stage four commences. `command.com` does its own initialization including creating the initial environment and executing `autoexec.bat`. Once this has occurred, `command.com` issues a prompt and goes into a loop awaiting user input. DOS-C is now loaded and ready to run.

I feel that it is always a good idea to include visual indicators in the code so that you can monitor the progress of an operation. When you boot DOS-C, the first indicator you will see is from boot. Because boot is small, it doesn't do very much except announce itself by placing the message "boot" on the screen. Although it doesn't seem like much, many times when I didn't see this little message, I immediately knew that I had an error to find.

`ipl.sys` announces that it is booting DOS-C, and the modified `.exe` loader outputs "." at regular intervals during the loading process. `ipl.sys` is bigger than boot and can afford to be somewhat more verbose. The reason for two visual indicators is twofold. First, you need an indicator that `ipl.sys` has been successfully loaded, the same as boot. Second, other things can go wrong with `ipl.sys`, so a visual indicator of the load process helps identify errors and the choice of putting a "." to the screen at strategic locations is very helpful. Finally, `kernel.exe` and `command.com` both announce themselves when loaded but do not contain progress indicators because the debugging technique used on them is different and gives better information than a visual indicator does.

Now that you have conceptually loaded DOS-C into memory successfully and have it operational, you can look a little closer at its organization. Next examine the design of the DOS-C kernel, and see how you can organize the code into logical modules, each implementing a logical system function.

DOS-C Kernel Architecture

The DOS-C architecture takes a layered approach. This approach is not very different from other operating systems. The only deviation you will find is that there is a decoupling between the upper and lower layers of the operating system.

Figure 3.2 shows the architecture of DOS-C. Because of the need for compatibility with the de facto DOS standards, full compliance is necessary in two interface areas. The first interface area is obvious — the DOS API entry points. These are the entry points that you expect from a DOS-compatible operating system. These include the traditional `int 20h, 21h, 22h, 23h, 24h, 27h, 28h,` and `2fh`.

Another less obvious interface is the device driver interface. This interface is also well documented and supported by many hardware and software vendors. It is important that DOS-C is compatible with this interface so that you can assure proper support of loadable device drivers.

The kernel consists of C code wherever possible. However, certain functions are written in assembly language. These functions typically perform direct machine control, such as manipulating the stack and flipping interrupt bits. You will find this code in a number of assembly files along with C-to-assembly interface functions.

Following a system call through the kernel is an easy way to take a tour through the kernel. Taking this journey and noting where certain features exist will help when you examine them more closely.

DOS-C API

A DOS API entry point handler intercepts any system call to DOS-C and translates it into a C call. Do this as close to the entry point as possible in order to use C as early as possible. The C call then does some context switching and performs a call to the appropriate internal service function. It is up to the service routine to route the system call through the kernel to the appropriate handler.

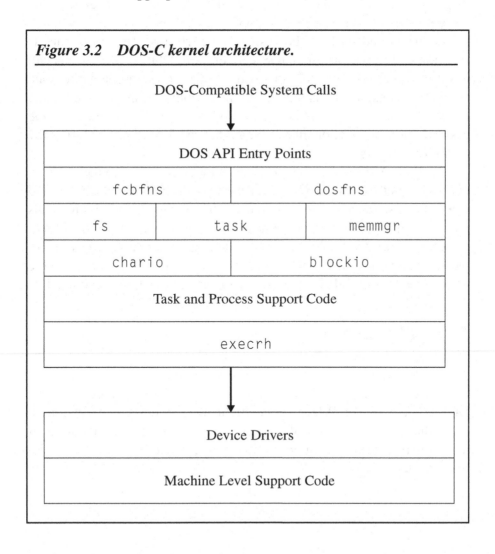

Figure 3.2   DOS-C kernel architecture.

DOS-Compatible System Calls

DOS API Entry Points

fcbfns	dosfns

fs	task	memmgr

chario	blockio

Task and Process Support Code

execrh

Device Drivers

Machine Level Support Code

The path through the kernel varies depending on which DOS API entry point you examine. In order to continue the search deeper into the kernel, focus on the int 21h service function, int21_service. This handler is a dispatcher that examines the call number found in the register storage area. It uses this number to execute some local machine-dependent code that converts register variables to C parameters for function calls, calls an internal function that performs the call, and converts the return values back into register values before returning. The dispatcher returns to the handler for another context switch and the entry point handler returns to the user.

As mentioned earlier, the dispatcher performs the appropriate C function call to perform the DOS-C operation. DOS-C directs this call into the kernel and can only proceed "down" in the architecture. One architecture design rule is that functions are only aware of functions at the same or lower layer. This is a DOS-C design choice that allows the easy addition of functionality at the higher levels without affecting the operational layers.

The DOS API layer directs a larger number of calls into the dosfns, or DOS Functions, layer. This layer works with the lower layers to add the DOS appearance to the system calls. It does this by combining calls into the various managers in the next lower layers with code to add features to the manager calls that are unique to a DOS interface. For example, this layer contains functions that encompass all rules for handling file I/O that deals with PSP, handles, or character and block I/O functionality.

DOS-C File System

The DOS-C design allows for a file manager that is independent from the design of the rest of the operating system. The fs module is the DOS-C file manager. Whenever an application requests an operation on a file that is resident on a block device, fs performs the operation. It does this by working with an internal table, called the f_node (or file node) table.

The f_node table is the foundation for the design of the FAT file system manager. The f_node data structure controls all internal file operations and virtualizes the file. When fs performs an operation such as create or open, it allocates an entry from the f_node table and populates the f_node fields. The call returns a number that is the index into the f_node table. From here on out, any function that performs an operation on the file receives this number. The f_node table entry that corresponds to this index controls all file parameters. It is the anchor that ties together the three components of a file discussed in Chapter 2.

If you are familiar with DOS internals from the undocumented internals literature, you will recognize the need to map an internal f_node to the SFT (System File Table) structure used by DOS. The need for this mapping is a result of the many programs that expect an SFT entry for a file. This is one of the many unfortunate results of having an operating system that is open to probing from user programs — you lose control of internal data structures. Changing any internal data structure may break a working application. It also complicates DOS-C design because these considerations will occur repeatedly and must be addressed in a manner that guarantees DOS compatibility.

For file-related issues, you handle this through the dosfns layer function. The module dosfns handles this type of mapping, which is part of its "DOS personality" responsibility. At the fs level, it does its work in an "OS-neutral" mode in order to localize unique features to the personality module. This design lends itself to portability but will complicate some DOS personality issues. Fortunately, the benefits far outweigh the complications.

DOS-C Memory Management

Memory management in DOS-C is an area that requires close emulation of the MS-DOS techniques. Because DOS-C is a real-mode operating system, the memory management techniques presented here are much simpler than those you will find if you examine more advanced operating systems.

In a manner similar to the MS-DOS memory management scheme, DOS-C manages memory through the use of a memory block linked list that uses the same arena header as the one described in Chapter 2.

The arena header in DOS-C is known as the MCB (Memory Control Block). As with MS-DOS, this data structure is in front of each allocated and free memory block. DOS-C uses the same "M" and "Z" signatures to identify the block type. Also, the segment value of the PSP of the program it belongs to, or a 0008h if it belongs to DOS-C, identifies the owner of the block in the same way. DOS-C uses all other entries in an identical manner. Again, many programs make use of this information, and DOS-C provides this same information for the sake of compatibility. Unlike fs where DOS personality splits between it and dosfns, memmgr contains all DOS personality.

DOS-C provides the same API calls as MS-DOS to manage the arenas. It provides an allocation function and a dispose function along with a call used to set allocation strategy. Additionally, a number of calls are used internally within DOS-C for initialization of the arena and validation of the integrity of the arena whenever memory blocks return to the free memory pool.

DOS-C Task Management

Because the DOS-C design adheres to the de facto DOS standard, it should not be surprising that the operating system is not multitasking. This is not entirely true as you will see in later chapters. There are really three tasks in DOS-C, the user task, the kernel task, and the driver task, but the user only sees a single task. The task manager, task, manages this single user task.

The primary function of the task manager is to act as a task loader, unlike other operating systems where task loading is incidental to managing the task. Also, there are only two relatively simple executable file types, .com and .exe. We have seen descriptions of both in Chapter 2, but we will examine the dynamics of loading the two types of tasks.

Both executable file types start from a single load entry. It is up to DOS-C to identify which file type it is. It does this by examining the first two bytes of the file. If it is an "MZ", then it is an .exe file and task invokes the .exe loader, otherwise task assumes it to be a .com binary image and invokes the .com loader.

Both loaders initially follow similar code. They both allocate memory from memmgr to place the environment strings. The differences begin when the actual file loading occurs. The .com loader merely allocates memory and begins loading the file into memory for a maximum of 64Kb. The .exe loader, on the other hand, computes the size required and then allocates memory. It then proceeds to load the image. When task completes loading the image, it does a seek to the relocation offset and does a segment fixup (necessary for the segmented architecture of the 80x86 family).

From this point, both loaders proceed to create the PSP and clone the file table. They both end by allocating and initializing the tasks' registers and executing the task if so requested. The only minor difference here is that the initial ss:sp and cs:ip are default values for a .com, whereas task computes them from the .exe file header for the .exe file.

The user program termination handler is the final task code to be examined. In our task module, program termination is handled in a straightforward manner by closing all open files, restoring the parent's information, and, if it is not a TSR, releasing the program's memory.

DOS-C High-Level I/O and Device Drivers

So far, you have followed the system call through the DOS-C API and examined the various managers that provide the requested service. If the request is an I/O request, the system call now proceeds deeper into the kernel to perform raw I/O. The manager does this by placing a call into the next layer — the raw I/O handlers.

There are two handlers in the raw I/O layer, chario and blockio. Although each handler uses different methods, both handlers are the primary interface between the device drivers and the remainder of the kernel. Both handlers perform all the necessary buffer management for both types of I/O, including line buffer management.

The I/O handlers are loosely based on the UNIX I/O model. As with UNIX, two types of I/O are defined. The first type is the character I/O type. This type of I/O appears as a stream of bytes to the kernel. The kernel either sequentially reads a byte into the kernel or writes from the kernel. There are also functions to read a buffer into memory and handle the familiar DOS line editing functions.

The block I/O interface provides functions to read and write a block of data to a block device — usually a disk. Each block corresponds to a sector and is actually a data structure in a block cache. When the kernel reads data from a disk, blockio reads the sector into a buffer and places it into a Least Recently Used (LRU) chain. When blockio needs a new buffer, it writes the tail of the list to disk, if needed, and returns that buffer. The buffer then becomes available for data transfer into the buffer. When blockio completes the data transfer, the buffer goes to the head of the LRU chain, indicating that it is the newest buffer. Management functions handle these operations for dirty buffer write back, as well as buffer fill. Buffer management functions also perform operations such as LRU management and flush.

As with the API, the device driver interface is well documented. It is important that DOS-C closely follow this convention. Because the interface is designed for an assembly language system, a special assembly language function interfaces all device drivers. This function, execrh(), accepts a request packet from the I/O handlers that contains a function number requested by the device driver. It handles the correct sequence of calls for the device driver to the strategy and interrupt entry points and returns the packet to the I/O handler. The function execrh() is a C call, allowing an easy assembly language interface.

The DOS-C device drivers are the bottom layer of the kernel. Like the remainder of the kernel, C is used in these device drivers also. The device drivers perform the necessary device interface between the kernel and the device itself. DOS-C has the same device drivers as MS-DOS and they perform the same functions as their MS-DOS counterparts.

Final Notes

The discussion of the DOS-C architecture covers about 90% of the kernel. MS-DOS defines a few functions that don't neatly fit the architecture. These functions, such as time management, National Language Support, etc., are handled individually in the layer that best supports them. Unfortunately, the design trade-offs forced special handling. As you will see in later chapters, the benefits of the layered design far outweigh the inconvenience of special functions.

## Chapter 4

DOS-C Kernel: File System Manager

Up to now, I have covered generic DOS and DOS-C basics. I have looked at the architecture of DOS-C and discussed its layered approach, design rules, and calling conventions. The method I used was a hypothetical trip through the kernel, and I examined the flow of a generic system call through the kernel and the interface layers.

This chapter examines the code in the kernel from the device interface layer up and starts by examining the device driver interface. This interface is an important piece of DOS-C, since all device driver calls use this interface. I will then look at the lower half of the file system managers, namely the block device and the character device interfaces. I will examine the data structures used in both and detail the design methodology behind each. Finally, I will tie these together by presenting the FAT file system code.

Device Driver Interface

The design of the code that performs the device driver interface is one of the critical areas in the kernel. In order to use standard DOS device drivers, the kernel must call the device driver in a manner similar to other DOS-compatible operating systems. This is necessary to guarantee compatibility with third-party device drivers (i.e., your mouse driver, some

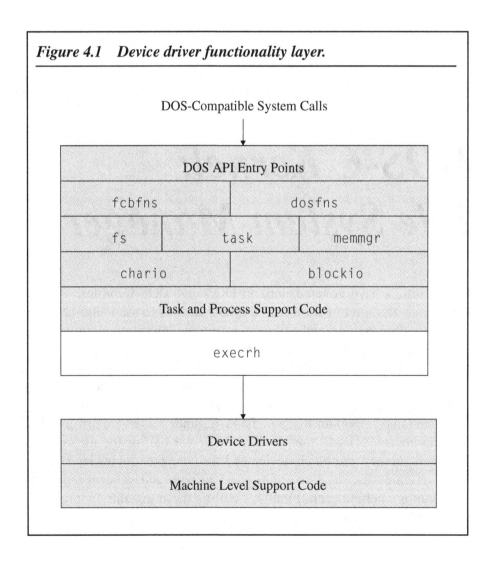

Figure 4.1   Device driver functionality layer.

SCSI disk drivers, etc.). It must also be easy to use by the kernel's C code modules. I achieve ease of use through a simple C function that receives a pointer to a standard DOS request packet and another pointer to a data structure, known as the dhdr, used to manage devices. This forms the device driver [execrh()] interface layer (Figure 4.1).

The realization of this design is an assembly code module that translates a C call into the appropriate device driver call. The file execrh.asm contains the actual assembly code; however, look at the pseudo code in Figure 4.2 in order to gain a better understanding of the process. The call takes two parameters, a far pointer, rhp, to the request packet data structure and another far pointer, dhp, to the dhdr data structure.

Figure 4.2   Psuedo code for execrh() *interface function.*

```
void
execrh(request far *rhp, struct dhdr far *dhp)
{
    void (*TempFunc)();

    /* perform c entry */
    *--sp = bp; /* Actually a push instruction*/
    bp = sp;
    sp -= 4;     /* and make space for TempFunc*/

    /* save old contents of temporary registers*/
    *--sp = si; /* Actually a push instruction*/
    *--sp = ds; /* ditto                */

    /*
     * Set up for standard DOS device driver call by assigning
     * passed in far pointers into their respective registers
     */
    ds:si = dhp;/* ds:es points to the device header */
    es:bx = rhp;/* es:bx points to the request packet */
```

The first parameter, rhp, is a pointer to the request packet data structure as described in various DOS references. The request packet is a variable data structure used by DOS to command a device to perform control, input, or output functions. The information contained within the request packet is all that is required by the device driver to perform the requested function. This request packet is the standard request packet expected by DOS-compatible device drivers.

Figure 4.2   Psuedo code for execrh() **interface function —**
continued.

```
/*
 * After translating C parameters to DOS standard
 * device driver parameters, begin executing driver
 * strategy and interrupt code by indirect calls
 * found in the device driver header.  Method used is
 * assigning the function pointer to a temporary
 * variable and using it for the function call.
 *
 * NOTE: device drivers use "tiny" model, so
 * register adjustments are made along the way.
 */

/* Call strategy first */
TempFunc = dhp -> dhStrategy;
TempFunc();
/* Call Interrupt next */
TempFunc = dhp -> dhInterrupt;
TempFunc();

/* and exit perform a c exit */
ds = *sp++; /* Actually a pop instruction*/
si = *sp++   /* ditto                    */
sp = bp;
}
```

In order to properly identify the device that performs the I/O request, execrh() requires a second parameter, dhp, the device header, which is loosely based on a similar data structure discussed in popular "undocumented" literature. It contains a pointer to the device driver variable and other useful information such as current directory for block devices.

Like other DOS-compatible operating systems, DOS-C creates a device I/O request in kernel space. Unlike some DOS-compatible systems, DOS-C creates this space as an automatic variable in the function that is performing I/O. DOS-C fills the packet then proceeds to call execrh() to execute the call.

Upon entry into the function, execrh() performs a sequence of instructions that emulate a normal C function entry, including reserving space for the local variable TempFunc. It next prepares for a standard DOS device driver call by assigning far pointers passed into execrh() to their respective registers — ds:es for the device header and es:bx for the request packet.

After translating the C parameters to DOS standard device driver parameters, execrh() continues by executing driver strategy and interrupt code with indirect calls found in the header of the device driver. The method used to perform the call is to load the function pointer into a temporary variable and use it for an indirect subroutine call. *Note:* Device drivers use the "tiny" model, so the actual assembly code makes register adjustments along the way.

Character Device Interface

The character device interface is one of the two lower-half functions used throughout the operating system (as diagrammed in Figure 4.3a, b). This interface handles both single-character I/O and buffered I/O, including command-line canonical processing. The file chario.c contains all functions related to this interface. With the understanding of how execrh() works, examine a simple function that uses it.

Let's examine the low-level character output function _sto() (List-ing 4.1). This function is responsible for all output to the system con-sole. The first thing it does is look for a break (Ctrl-C or Ctrl-Break). I will discuss the exact mechanism later, but for now just assume the existence of a function that returns TRUE if a user has hit a Break key. If a break condition exists, _sto() handles it and exits immediately; if not, it proceeds to read the character. As mentioned before, the request packet is actually an automatic variable allocated on the kernel stack; _sto() is a perfect example. The function fills the fields of the request, including the command requested. Each I/O function has different fields, but the concept is universal to all the low-level drivers.

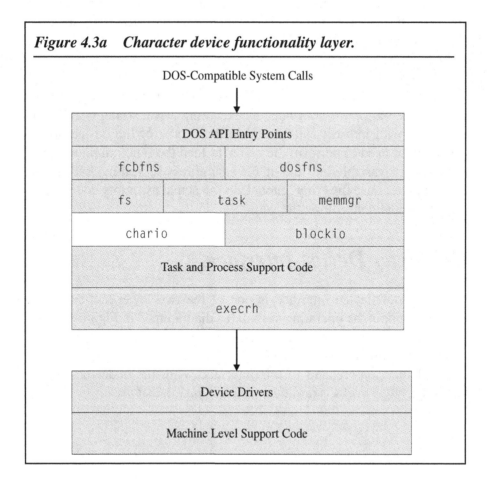

Figure 4.3a   Character device functionality layer.

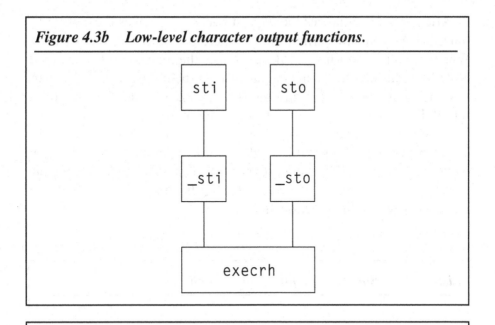

Figure 4.3b   Low-level character output functions.

Listing 4.1   Source code for _sto( ) function.

```
static BOOL
_sto(COUNT c)
{
    request rq;
    BYTE buf = c;

    if(con_break())
    {
        handle_break();
        return FALSE;
    }
    rq.r_length = sizeof(request);
    rq.r_command = C_OUTPUT;
    rq.r_count = 1;
    rq.r_trans = (BYTE FAR *)(&buf);
    rq.r_status = 0;
    execrh((request FAR *)&rq, (struct dhdr FAR *)&con_dev);
    if(rq.r_status & S_ERROR)
        return char_error(&rq, con_name);
    return TRUE;
}
```

After _sto( ) initializes the request packet, it invokes the execrh( ) function. As you saw earlier, this is where the device driver is called. When execrh( ) returns, _sto( ) examines the request packet status. If an error occurred, the kernel character error handler is invoked; otherwise, the function successfully returns. Again, this handling is typical for all the low-level device driver interface functions.

A function similar to _sto( ) is the sto( ) function (Listing 4.2). The function _sto( ) provides only primitive functionality. Users of modern operating systems, have become accustomed to seeing tab expansion, screen hold (Ctrl-S and Ctrl-Q handling), etc. The sto( ) function provides this functionality.

Listing 4.2   Source code for sto( ) function.

```
VOID
sto(COUNT c)
{
    /* Test for hold char    */
    con_hold();

    /* Display a printable character*/
    if(c != HT)
        _sto(c);
    if(c == CR)
        scr_pos = 0;
    else if(c == BS)
    {
        if(scr_pos > 0)
            --scr_pos;
    }
    else if(c == HT)
    {
        do
            _sto(' ');
        while(++scr_pos & 7);
    }
    else if(c != LF && c != BS)
        ++scr_pos;
}
```

The first order of business for sto() is to check for the reception of a hold character (Ctrl-S). This is done through the special function con_hold(), which pauses until a Ctrl-Q is received. As with other parts of DOS-C, sto() also builds on lower level functions, in this case _sto(), to achieve its functionality. sto() tests for special characters such as a horizontal tab (HT), enter, or carriage return (CR). Tab characters are expanded to multiples of eight spaces, whereas carriage return and backspace characters are intercepted by sto(), which modifies the global variable scr_pos, which tracks the current column position of the cursor. scr_pos is also used later in the command line editing functions.

Similar to _sto() is _sti() (Listing 4.3). Like _sto(), _sti() is a lower level function that performs input. Two levels of character I/O are needed because DOS-compatible operating systems can supply both raw (no tab expansions, backspace characters that do not delete previous characters, etc.) and processed or "cooked" (tab expansions, backspace

---

Listing 4.3   Source code for _sti() function.

```
COUNT
_sti(VOID)
{
    BYTE cb;
    request rq;

    rq.r_length = sizeof(request);
    rq.r_command = C_INPUT;
    rq.r_count = 1;
    rq.r_trans = (BYTE FAR *)&cb;
    rq.r_status = 0;
    execrh((request FAR *)&rq, (struct dhdr FAR *)&con_dev);
    if(rq.r_status & S_ERROR)
        return char_error(&rq, con_name);
    if(cb == CTL_C)
    {
        handle_break();
        return CTL_C;
    }
    else
        return cb;
}
```

characters that delete previous characters, etc.) I/O. Because both input and output require the two levels of support, a similar design is used in both _sto() and _sti().

The function sti() supplies the next layer of functionality (Listing 4.4). Here, the symmetry breaks down because the DOS compatibility rule dictates that the next level fills a special keyboard buffer and performs the expected line editing. As a result, sti() is much longer that sto().

sti() begins by checking the line buffer that is passed in to see if it is large enough to hold the requested number of bytes and for a carriage return terminating the buffer if it is not empty. The higher layer read functions use this particular feature. (The preprocessor switch NOSPCL is discussed later).

sti() then falls into a loop that terminates only when _sti() returns a carriage return. At the beginning of the loop, sti() then tests a character with the use of a switch. The switch defaults to a simple entry of the character into the keyboard buffer with the use of a helper function, kbfill(). The switch captures other characters, such as backspace and escape, and diverts them to special handling code. It is this mechanism that produces the familiar repeated line response to the F3 function key and the erased character with both the backspace key and the back-arrow key.

One interesting note here is the use of a static character array as a buffer for the last line. As with all DOS operating systems, DOS-C provides the ability to recall the last line entered a character at a time with the use of the right-arrow key. When DOS-C receives a carriage return, it transfers the line into the static buffer local_buffer in order to properly perform line editing. DOS-C replaces the received character by the character contained in local_buffer at the correct position within the line as recorded by structure member, kp -> kb_count. This buffer is also used to repeat the entire last line whenever the F3 function key is received. I must warn you that this particular design is not meant for multiprogramming environments because the use of a static variable defeats the reentrant requirement that all functions must follow in a true multiprogramming kernel. However, DOS-C is not multiprogramming, so you can allow some liberty and use simpler code in the function.

Listing 4.4   Source code for `sti()` function.

```
VOID
sti(keyboard FAR *kp)
{
    REG UWORD c, cu_pos = scr_pos;
    WORD init_count = kp -> kb_count;
#ifndef NOSPCL
    static BYTE local_buffer[LINESIZE];
#endif

    if(kp -> kb_size == 0)
        return;
    if(kp -> kb_size <= kp -> kb_count || kp -> kb_buf[kp ->
kb_count]
        != CR)
        kp -> kb_count = 0;
    FOREVER
    {
        switch(c = _sti())
        {
        case CTL_F:
            continue;

#ifndef NOSPCL
        case SPCL:
            switch(c = _sti())
            {
            case LEFT:
                goto backspace;

            case F3:
            {
                REG COUNT i;

                for(i = kp -> kb_count; local_buffer[i] != '\0';
i++)
                {
                    c = local_buffer[kp -> kb_count];
                    if(c == '\r' || c == '\n')
                        break;
                    kbfill(kp, c, FALSE);
                }
                break;
```

Listing 4.4   Source code for `sti()` ***function — continued.***

```
            case RIGHT:
                c = local_buffer[kp -> kb_count];
                if(c == '\r' || c == '\n')
                    break;
                kbfill(kp, c, FALSE);
                break;
            }
        break;
#endif

        case CTL_BS:
        case BS:
        backspace:
            if(kp -> kb_count > 0)
            {
                if(kp -> kb_buf[kp -> kb_count - 1] >= ' ')
                {
                    destr_bs();
                }
                else if((kp -> kb_buf[kp -> kb_count -1] < ' ')
                        && (kp -> kb_buf[kp -> kb_count - 1]!= HT))
                {
                    destr_bs();
                    destr_bs();
                }
                else if(kp -> kb_buf[kp -> kb_count -1] == HT)
                {
                    do
                    {
                        destr_bs();
                    }
                    while((scr_pos > cu_pos) &&  (scr_pos & 7));
                }
                --kp -> kb_count;
            }
            break;

        case CR:
            kbfill(kp, CR, TRUE);
            kbfill(kp, LF, TRUE);
#ifndef NOSPCL
            fbcopy((BYTE FAR *)kp -> kb_buf, (BYTE FAR *)local_buffer,
                (COUNT)kp -> kb_count);
            local_buffer[kp -> kb_count] = '\0';
#endif
```

The file chario.c also contains a number of helper functions that functions in chario.c and other files use. These functions perform tasks, such as converting displayed control characters into a printable form by placing a "^" in front of the corresponding printable character [mod_sto()]. Other helper functions fill the keyboard buffer and prevent buffer overrun conditions while warning the user with an audible alert [kbfill()]. I suggest that you study these functions in order to get a better understanding of the interrelationships between the main I/O functions and the helper functions.

One final note about sti(). The DOS-C design allows for non-IBM-compatible I/O, such as a serial interface, where these function keys (F3, right arrow, and left arrow) may not exist. In order to work in this environment, you must define NOSPCL during the compilation of chario.c. This removes the local_buffer functionality from the function sti().

Listing 4.4   Source code for sti() **function — continued.**

```
        return;

    case LF:
        sto(CR);
        sto(LF);
        break;

    case ESC:
        sto('\\');
        sto(CR);
        sto(LF);
        for(c = 0; c < cu_pos; c++)
            sto(' ');
        kp -> kb_count = init_count;
        break;

    default:
        kbfill(kp, c, FALSE);
        break;
    }
  }
}
```

Block Device Interface

The block device interface is similar to the character device design. Like its character counterpart, the block device interface has a lower layer that acts as the device driver interface and an upper layer that performs higher level functions such as buffering (Figure 4.4a,b). However, they both differ due to the nature of the I/O.

At the API level, any file or device can have a read or write request of an arbitrary number of bytes. For the character device interface, this

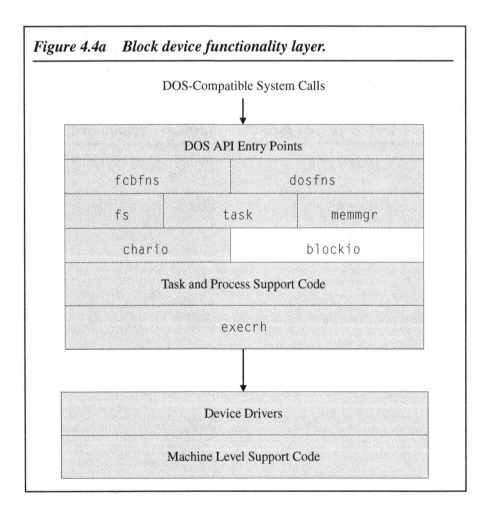

Figure 4.4a   Block device functionality layer.

DOS-Compatible System Calls

DOS API Entry Points

| fcbfns | dosfns |

| fs | task | memmgr |

| chario | blockio |

Task and Process Support Code

execrh

Device Drivers

Machine Level Support Code

is usually not a problem because character devices are capable of transferring a single byte at a time. Block I/O devices must transfer data a
fixed number of bytes at a time, 512 bytes for a DOS-compatible operating system. Also, block devices are random access, meaning that any
block can be accessed in any order, with finite rotational latency and
cylinder stepping times incurred with each read from a disk block.
Therefore, our lower layer for the design of the block device interface
buffers these blocks in a way that reduces disk access times. DOS-C
uses a linked list of block buffer structures that contains flags, pointers,
and a data buffer. A last-recently-used list updated with every buffer
access organizes the buffer structures. DOS-C reads the buffers into
memory only when it needs a buffer not in the cache and writes a buffer
to disk when it needs a new buffer that is not in the cache and the buffer
removed from the end of the list was modified. DOS-C also writes to
disk when it specifically flushes the disk buffers.

Similar to the character I/O functions, the block I/O functions also
have a device driver interface that converts the assembly language calling conventions to C. For our block I/O functions, the call is
dskxfer() (Listing 4.5).

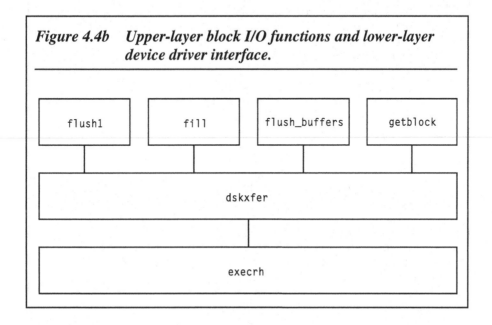

**Figure 4.4b   Upper-layer block I/O functions and lower-layer
device driver interface.**

On first examination of dskxfer(), the system call does not seem very different from its character counterpart except that the set-up of the request packet is inside an infinite loop. I will address the reason for this shortly, but it will be an important item to note.

When you look closer at the set-up in dskxfer(), you see that the block unit passes a subunit. The use of a subunit allows a device driver to handle multiple drives (i.e., the floppy driver needs to handle both A and B drives while the hard disk driver handles C and D drives). You also see that the command can be any one of three: C_OUTVFY, C_OUTPUT, and

---

Listing 4.5   Source code for *dskxfer( )* function.

```
BOOL
dskxfer (COUNT dsk, LONG blkno, VOID FAR *buf, COUNT mode)
{
    REG struct dpb *dpbp = &blk_devices[dsk];
    request rq;

    for(;;)
    {
        rq.r_length = sizeof(request);
        rq.r_unit = dpbp -> dpb_subunit;
        rq.r_command =
         mode == DSKWRITE ?
           (verify_ena ? C_OUTVFY : C_OUTPUT)
            : C_INPUT;
        rq.r_status = 0;
        rq.r_meddesc = dpbp -> dpb_mdb;
        rq.r_trans = (BYTE FAR *)buf;
        rq.r_count = 1;
        if(blkno > MAXSHORT)
        {
            rq.r_start = HUGECOUNT;
            rq.r_huge = blkno - 1;
        }
        else
            rq.r_start = blkno - 1;
        execrh((request FAR *)&rq, dpbp -> dpb_device);
        if(!(rq.r_status & S_ERROR) && (rq.r_status & S_DONE))
            break;
```

C_INPUT. As you can see, dskxfer() addresses both input and output. Additionally, there are two output modes, a simple write mode (C_OUTPUT) and a write then verify mode (C_OUTVFY). At one time, disk hardware and media were not as reliable as they are today, so it was foolish to operate without this mode. Today's hardware is much more reliable, so you don't really need to operate in this mode. The C_OUTVFY mode is also time consuming, requiring that the write operation complete and the disk spin one full revolution to perform the verify operation.

The next area in which you will notice a difference is where dskxfer() passes the block number, blkno, into the request packet. You will notice that when the block number is greater than MAXSHORT, dskxfer() sets the r_start field to HUGECOUNT and the r_huge field to blkno - 1; otherwise, dskxfer() sets the r_start field to blkno - 1.

Listing 4.5   Source code for dskxfer() function — continued.

```
        else
        {
        loop:
            switch(block_error(&rq, dpbp -> dpb_unit))
            {
            case ABORT:
            case FAIL:
                return FALSE;

            case RETRY:
                continue;

            case CONTINUE:
                break;

            default:
                goto loop;
            }
        }
    }
    return TRUE;
}
```

The two cases are handled similarly because both require that blkno be decremented, although the internal representation of a block device starts with one, whereas the device driver starts with zero. Two data structure members are needed. Because DOS now allows more than 65,536 blocks to a device. When the requested block exceeds 65,535, a 32-bit field is used instead of the traditional 16-bit field. The need for 32-bit block addresses resulted from the availability of drives with greater than 32Mb (65,535*512 bytes). The use of a split handler allows for backward compatibility with older disk drivers.

After dskxfer() fills the request packet, it makes a device driver call through the same execrh() function that the character device uses. It is after this call that the function differs from its character counterpart. Where the corresponding character device handler simply exits through the character error handler, dskxfer() has a loop. The loop starts out by invoking the block error handler, the function that outputs the familiar "Abort, retry, fail" error message. Depending on what the error handler returns (presumably, the user responded with one of the three choices), the loop is responsible for the resulting action. For a retry response, the loop forces the whole operation to be retried through a simple C continue statement. For a fail or abort response, the loop forces an error return. For a continue response, the loop allows DOS-C to ignore the error altogether: a C break statement allows the function to exit through the normal function return so that a success is returned regardless of what the driver reported. Note that the error handler must return a valid condition; otherwise, dskxfer() repeats the loop until it receives a known return code. The use of the infinite loop forces the process to repeat until the user returns a valid response.

The next step up from the disk driver interface dskxfer() is flush1() (Listing 4.6). This function moves up one level by working with the block structure, unlike dskxfer() which works with the data itself. This allows flush1() to work more efficiently through key block structure members and also gives it full control over when to do a physical write.

The first order of business for flush1() is to examine the members b_flag, which indicates that the data in the buffer is valid if TRUE, and b_update, which indicates that the data in the buffer has been modified. If both these members are true, flush1() transfers the data to disk. When flush1() completes the operation, it clears b_update, indicating the data has been saved to disk, and clears b_flag if an error occurred during the write process.

Like its companion flush1(), fill() also is in the next layer up from the device driver interface (Listing 4.7). Where flush1() handles the write operation, fill() handles the read operation. fill() is similar to flush1() with one major exception: it must save the buffer, if it was modified to disk. In order to do this, the entry code is different.

Upon entry, fill() examines the member b_flag, which indicates whether or not it contains valid data. It also examines the member b_update, which records whether the data has been modified. If both conditions are true, fill() calls flush1() to clear the buffer before proceeding any further. Next, it performs a precautionary test to ensure

---

Listing 4.6   Source code for `flush1()` ***function.***

```
BOOL
flush1 (struct buffer FAR *lpBlock)
{
    REG WORD ok;

    if (lpBlock -> b_flag && lpBlock -> b_update)
        ok = dskxfer(lpBlock -> b_unit, lpBlock -> b_blkno,
            (VOID FAR *)lpBlock -> b _buffer, DSKWRITE);
    else
        ok = TRUE;
    lpBlock -> b_update = FALSE;     /* even if error,        */
                                     /* mark not dirty        */
    lpBlock -> b_flag &= ok;         /* otherwise system has  */
                                     /* trouble continuing.   */
    return(ok);
}
```

that if the buffer did require a write operation, it completed properly. fill() accomplishes this through the local variable ok, which contains the return code if a write operation was performed. If the operation does not require a write, fill() sets ok to TRUE in order to guarantee the read operation takes place.

Once the dskxfer() operation completes, fill() assigns the return code to the member b_flag. This invalidates the buffer if the read operation failed. Finally, fill() clears the member b_update, indicating that the buffer is clean, and updates the members b_blkno and b_unit to identify the disk block associated with this buffer.

With the single buffer operations in place, the next step up is the buffer chain operations. These operations are the ones that maintain the chain and implement the LRU functionality. There is really only one active chain entry point, getblock(), but there are a number of chain

---

Listing 4.7   Source code for *fill()* function.

```
BOOL
fill (REG struct buffer FAR *lpBlock, LONG blkno, COUNT dsk)
{
    REG WORD ok;

    if(lpBlock -> b_flag && lpBlock -> b_update)
        ok = flush1(lpBlock);
    else
        ok = TRUE;

    if(ok)
        ok = dskxfer(dsk, blkno, (VOID FAR *)lpBlock -> b_buffer,
                    DSKREAD);
    lpBlock -> b_flag = ok;
    lpBlock -> b_update = FALSE;
    lpBlock -> b_blkno = blkno;
    lpBlock -> b_unit = dsk;
    return(ok);
}
```

maintenance functions. The function flush_buffers() is an example of a chain maintenance function (Listing 4.8).

Sometimes DOS-C needs to ensure that all the disk buffers are written to disk. This condition may occur when all files on a disk are closed and files were written to. This is an example of when DOS-C calls flush_buffers().

The function is simple. It begins by initializing the variable lpBlock. It uses a while loop to walk the chain, examining each buffer on the chain for a match to the requested disk. If the buffer belongs to the requested disk, flush_buffers() invokes flush1() to write the buffer out. It then picks up the next buffer address from the b_next member and continues the loop. Note that the loop terminates with a null pointer that terminates the chain.

Other simple functions perform operations similar to flush_buffers(), such as init_buffers(), which initializes the chain; flush(), which flushes all buffers to disk; and setinvld(), which invalidates a number

---

Listing 4.8   *Source code for* flush_buffers() *function.*

```
BOOL
flush_buffers (REG COUNT dsk)
{
    REG struct buffer FAR *lpBlock;
    REG BOOL ok = TRUE;

    lpBlock = firstbuf;
    while (lpBlock)
    {
        if(lpBlock -> b_unit == dsk)
            if(!flush1(lpBlock))
                ok = FALSE;
        lpBlock = lpBlock -> b_next;
    }
    return ok;
}
```

of buffers on the chain. These are chain maintenance functions whose need arises from various DOS-C conditions, but a single entry, as mentioned earlier, performs the LRU maintenance. This results from the fundamental design rule for block I/O. DOS-C reads and writes all disk blocks through the buffer chain. This simplifies stream I/O at the higher levels. All higher level functions request a block and modify it. When it becomes necessary to update a block, DOS-C performs all updates through the buffer chain maintenance functions. As a result, there are no explicit read or write block calls.

Listing 4.9   Source code for `getblock()` **function.**

```
struct buffer FAR *
getblock (LONG blkno, COUNT dsk)
{
    REG struct buffer FAR *lpBlock;
    REG struct buffer FAR *lpLastBlock;
    REG struct buffer FAR *lpMiddleBlock;
    REG WORD imsave;

    /* Search through buffers to see if the required block   */
    /* is already in a buffer                                */

    lpBlock = firstbuf;
    lpLastBlock = NULL;
    lpMiddleBlock = NULL;
    while(lpBlock != NULL)
    {
        if ((lpBlock -> b_flag) && (lpBlock -> b_unit == dsk)
            && (lpBlock -> b_blkno == blkno) )
        {
            /* found it -- rearrange LRU links               */
            if(lpLastBlock != NULL)
            {
                lpLastBlock -> b_next = lpBlock -> b_next;
                lpBlock -> b_next  = firstbuf;
                firstbuf = lpBlock;
            }
            return(lpBlock);
        }
```

The function `getblock()` is responsible for returning a block to any function that needs it (Listing 4.9). In order to do this and follow the basic design rules, the design for `getblock()` is in two parts. First it searches the chain, and next it removes a block. How it removes the block is what determines the read or write operation.

The function starts off by initializing three pointers: `lpBlock`, `lpLastBlock`, and `lpMiddleBlock`. The pointer `lpBlock` points to the current block and is the variable used to terminate the `while` loop following the initialization. Three conditions determine if a block is found: (1) the block is valid (`b_flag`), (2) it is from the correct disk (`b_unit`), and (3) it is the block we want (`b_blkno`). If these conditions are met, `getblock()` removes the buffer from the chain and places it at the front of the chain before returning the buffer address; otherwise, it

---

Listing 4.9   Source code for `getblock()` function — continued.

```
        else
        {
            /* move along to next buffer*/
            lpMiddleBlock = lpLastBlock;
            lpLastBlock = lpBlock;
            lpBlock  = lpBlock -> b_next;
        }
    }
    /* The block we need is not in a buffer, we must make a    */
    /* buffer available, and fill it with the desired block    */

    /* detach lru buffer */
    if(lpMiddleBlock != NULL)
        lpMiddleBlock -> b_next = NULL;
    lpLastBlock -> b_next = firstbuf;
    firstbuf = lpLastBlock;
    if(flush1(lpLastBlock) && fill(lpLastBlock, blkno, dsk))
    /* success */
        lpMiddleBlock = lpLastBlock;
    else
        lpMiddleBlock = NULL;    /* failure                    */
    return (lpMiddleBlock);
}
```

reassigns the pointers and moves the current buffer pointer to the next buffer in the chain. If the `while` loop terminates, then the buffer is not on the chain and `getblock()` removes the last buffer from the chain. `flush1()` writes out the block's contents, if necessary, and `fill()` reads the desired block's contents. `getblock()` then returns the block address to the caller, conditioned by error handling.

FAT File System Functions

Three files make up the `fs` module: `fattab.c`, `fatdir.c`, and `fatfs.c`. I chose to partition the functions that make up `fs` in this way so that each of these three files corresponds to the three distinct objects that make up the FAT file system disk discussed in Chapter 2. The file `fattab.c` deals with FAT handling on a device. It works with both 12-bit and 16-bit FAT devices, and it treats both in a uniform fashion. The file `fatdir.c` handles all directory management and is responsible for directory searches (i.e., Find First and Find Next). The file `fatfs.c` handles all the remaining data functions and maps these functions into a support subset for the `dosfns` and `fcbfns` personality layers.

FAT Management

As you have seen before, a symmetry similar to the read and write symmetry seen in both the character and block device interfaces exists for both operations of linking a FAT block number into a chain and retrieving a FAT block number from the FAT chain (Figure 4.5). The file `fattab.c` is made up of a FAT block retrieval portion (read) and a FAT block linking portion (write). The retrieval portion entry is `next_cluster()`. This function returns the next cluster in a chain given the current cluster. Similarly, the linking portion entry is `link_fat()`. This function takes two FAT cluster numbers and links the second cluster into the first cluster.

Retrieving a FAT Cluster Number

The best way to look at these functions is to start with next_cluster()
and then examine link_fat(). Because next_cluster() does not
modify any data structures, I will use it to study the fundamental algorithms. I will proceed by looking at the 16-bit handlers, then the 12-bit
handlers. The 16-bit handlers are easier to understand because they
don't have to deal with packing issues resulting from the mapping of
12-bit values into byte-oriented data structures. Using this approach,
you get an opportunity to look at the FAT closely before dealing with the
complications of the linking functions.

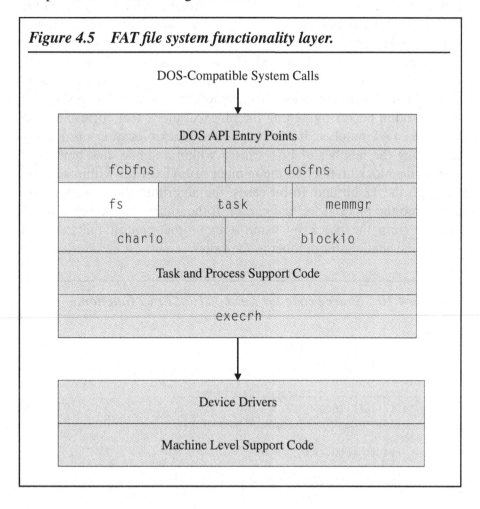

Figure 4.5   FAT file system functionality layer.

The design of the upper layer of the FAT file system requires that the FAT handlers treat all disks in a uniform manner. As discussed earlier, there are two different FAT sizes: 12-bit and 16-bit. Because the higher level functions don't concern themselves with the type of FAT, only a single entry point is provided: next_cluster() (Listing 4.10). The requirements of this function are relatively simple: it must decide what type of FAT the disk uses and then dispatch the correct handler. The macro ISFAT12 (Figure 4.6) is used to decide which function to call by first testing the disk parameter block dpb_size member to determine what size FAT the current device contains. This test looks to see if the size of the disk is less than or equal to FAT_MAGIC (4086) blocks. If it is, then the FAT is a 12-bit type and next_cluster() invokes next_cl12(); otherwise it invokes next_cl16().

The two FAT handlers are both similar in design, but I will examine next_cl16() first (Listing 4.11). This provides an opportunity to look cleanly at the algorithm that walks a linked list in the FAT. In each case, the algorithm begins by getting the physical block that corresponds to the given FAT number. It determines the blocks location on disk by computing the number of FAT entries within a single disk block and scaling the block number by the number of FAT entries within a block. For both the 12-bit and 16-bit cases, the algorithm then computes an index into the block it fetched earlier. It then retrieves the next cluster number from the block by using a byte-order-neutral function and returns the next cluster number to the caller.

Listing 4.10   Source code for next_cluster() ***function.***

```
UWORD
next_cluster (struct dpb *dpbp, REG UCOUNT ClusterNum)
{
    if(ISFAT12(dpbp))
        return next_cl12(dpbp, ClusterNum);
    else if(ISFAT16(dpbp))
        return next_cl16(dpbp, ClusterNum);
    else
        return LONG_LAST_CLUSTER;
}
```

Figure 4.6 *ISFAT12 macro for use with* `next_cluster()`.

```
#define ISFAT12(dpbp) (((dpbp)->dpb_size)<=FAT_MAGIC)

     .
     .
     .


WORD
next_cluster (struct dpb *dpbp, REG UCOUNT ClusterNum)
{
 if(ISFAT12(dpbp))
  return next_cl12(dpbp, ClusterNum);
 else if(ISFAT16(dpbp))
  return next_cl16(dpbp, ClusterNum);
 else
  return LONG_LAST_CLUSTER;
}
```

Listing 4.11 *Source code for* `next_cl16()` *function.*

```
UWORD
next_cl16 (struct dpb *dpbp, REG UCOUNT ClusterNum)
{
    UCOUNT idx;
    struct buffer FAR *bp;
    UWORD RetCluster;

    /* Get the block that this cluster is in */
    bp = getblock((LONG)(((LONG)ClusterNum) * SIZEOF_CLST16) /
                  dpbp -> dpb_secsize + dpbp ->
                  dpb_fatstrt + 1,dpbp -> dpb_unit);
    if(bp == NULL)
        return DE_BLKINVLD;

    /* form an index so that we can read the block as a     */
    /* byte array                                           */
    idx = (((LONG)ClusterNum) * SIZEOF_CLST16) %
          dpbp -> dpb_secsize;

    /* Get the cluster number, */
    fgetword((VOID FAR *)&(bp -> b_buffer[idx]),
             (WORD FAR *)&RetCluster);

    /* and return successful. */
    return RetCluster;
}
```

The differences between the 12-bit and 16-bit cases become apparent when you compare next_cl12() to next_cl16(). next_cl12() is shown in Listing 4.12. The most obvious difference is in size. next_cl12() is larger than its 16-bit counterpart, which results from the special unpacking that it must perform. The 12-bit FAT packs two FAT numbers into 3 bytes, saving 1 byte for every two entries when compared to the 16-bit FAT. However, the complexity of this operation forces next_cl12() to work harder.

Like its 16-bit counterpart, next_cl12() begins by getting the block that corresponds to the given FAT cluster number. It also determines the block by computing the number of FAT cluster entries within

---

Listing 4.12   Source code for *next_cl12()* function.

```
UWORD
next_cl12 (struct dpb *dpbp, REG UCOUNT ClusterNum)
{
    REG UBYTE FAR *fbp0, FAR *fbp1;
    UCOUNT idx;
    struct buffer FAR *bp, FAR *bp1;

    /* Get the block that this cluster is in */
    bp = getblock((LONG)(((((ClusterNum << 1) + ClusterNum) >> 1) /
                        dpbp -> dpb_secsize + dpbp ->
                        dpb_fatstrt + 1), dpbp -> dpb_unit);
    if(bp == NULL)
        return BAD;

    /* form an index so that we can read the block as a        */
    /* byte array                                              */
    idx = (((ClusterNum << 1) + ClusterNum) >> 1) % dpbp -> dpb_secsize;

    /* Test to see if the cluster straddles the block. If it   */
    /* does, get the next block and use both to form the       */
    /* FAT word. Otherwise, just point to the next block.      */
    if(idx >= dpbp -> dpb_secsize - 1)
    {
        bp1 = getblock((LONG)(dpbp -> dpb_fatstrt +
                    (((((ClusterNum << 1) + ClusterNum) >> 1) /
                    dpbp -> dpb_secsize)) + 2, dpbp -> dpb_unit);
        if(bp1 == (struct buffer *)0)
            return BAD;
        fbp1 = (UBYTE FAR *)&(bp1 -> b_buffer[0]);
    }
```

a single disk block and scaling it. At this point, the 12-bit algorithm diverges from the 16-bit algorithm. The 12-bit FAT table entry always straddles 2 bytes and the algorithm computes an index into the block. However, because of the 3-to-2 ratio of bytes-to-entries, certain block sizes may cause the last entry in the block to straddle into the next block. For example, in a 512-byte block, 341 entries occupy 511 bytes, leaving a single byte in each 512-byte block. This results in the unique situation that if the index points to the last byte in the block, the entry straddles the next block.

At this point, next_cl12() determines if the FAT cluster number straddles the block with the use of two pointers to build the FAT cluster number. If the entry starts on the last byte of the block, next_cl12() reads the next block and sets the second pointer to the first byte, otherwise it simply sets the second pointer to the next byte in the block.

At this point, next_cl12() fetches the 2 bytes that compose the next cluster number from the block or blocks by using a special method that differentiates odd-numbered from even-numbered FAT entries, which are packed differently. next_cl12() performs a test for odd-numbered FAT entries and executes the correct unpacking procedure

---

Listing 4.12   Source code for next_cl12() **function —**
                continued.

```
    else
        fbp1 = (UBYTE FAR *)&(bp -> b_buffer[idx + 1]);
    fbp0 = (UBYTE FAR *)&(bp -> b_buffer[idx]);
    /* Now to unpack the contents of the FAT entry. Odd and        */
    /* even bytes are packed differently.                          */
    if (ClusterNum & 0x01)
        ClusterNum = ((*fbp0 & 0xf0) >> 4) | *fbp1 << 4;
    else
        ClusterNum = *fbp0 | ((*fbp1 & 0x0f) << 8);
    if ((ClusterNum & MASK) == MASK)
        ClusterNum = LAST_CLUSTER;
    else if ((ClusterNum & BAD) == BAD)
        ClusterNum = BAD;
    return ClusterNum;
```

based on the outcome. The function returns this number unless it encounters a last cluster or bad cluster value, in which case it returns the 16-bit equivalent instead, satisfying the FAT size independence requirement of the upper layers.

Writing a FAT Cluster Number

As with the next_cluster() function, the higher level functions don't concern themselves with the type of FAT, so only a single entry point is provided: link_fat() (Listing 4.13). The requirements of this function are the same as those of next_cluster(): it must decide what type of FAT the disk uses and then dispatch the correct handler. It performs these tasks in exactly the same way as next_cluster() and dispatches either link_fat12() or link_fat16() for 12-bit and 16-bit FATs respectively.

Again the two FAT handlers are both similar in design, and I will examine the 16-bit handler, link_fat16(), first (Listing 4.14). As you already know, the 16-bit version is simpler than the 12-bit version because it doesn't concern itself with packing and unpacking cluster numbers.

As with the next_cluster() functions, the link_fat() algorithms begin by getting the block that corresponds to the given FAT number. They both determine the block by computing and scaling the number of FAT entries within a single disk block. For both the 12-bit and 16-bit cases, the algorithm then computes an index into the block and places

---

Listing 4.13   Source code for link_fat() **function.**

```
UCOUNT
link_fat (struct dpb *dpbp, UCOUNT Cluster1, REG UCOUNT Cluster2)
{
    if(ISFAT12(dpbp))
        return link_fat12(dpbp, Cluster1, Cluster2);
    else if(ISFAT16(dpbp))
        return link_fat16(dpbp, Cluster1, Cluster2);
    else
        return DE_BLKINVLD;
}
```

the next cluster number into the block using a byte-order-neutral function. Unlike the `next_cl16()` function, `link_fat16()` must now mark the block as modified in order to guarantee that the FAT is properly updated. It also updates the dpb, which is an internal data structure

---

Listing 4.14   Source code for `link_fat16()` ***function.***

```
UCOUNT
link_fat16 (struct dpb *dpbp, UCOUNT Cluster1, UCOUNT Cluster2)
{
    UCOUNT idx;
    struct buffer FAR *bp;
    UWORD C12 = Cluster2;

    /* Get the block that this cluster is in */
    bp = getblock((LONG)(((LONG)Cluster1) * SIZEOF_CLST16) /
                dpbp -> dpb_secsize + dpbp -> dpb_fatstrt + 1,
                dpbp -> dpb_unit);
    if(bp == NULL)

    /* Finally, put the word into the buffer and mark the      */
    /* buffer as dirty.                                         */
    fputword((WORD FAR *)&C12, (VOID FAR *)&(bp ->
b_buffer[idx]));
    bp -> b_update = TRUE;

    /* Return successful.                                       */
    /* update the free space count                             */
    if(Cluster2 == FREE)
    {
        /* update the free space count for returned cluster    */
        if(dpbp -> dpb_nfreeclst != UNKNCLUSTER)
            ++dpbp -> dpb_nfreeclst;
    }
    else
    {
        /* update the free space count for removed cluster     */
        if(dpbp -> dpb_nfreeclst != UNKNCLUSTER)
            --dpbp -> dpb_nfreeclst;
    }

    return SUCCESS;
}
```

associated with the disk. It increases the dpb_nfreeclst member if the linked cluster is actually a free cluster entry and decreases it otherwise. This way, the structure is updated as we remove clusters from and return clusters to the free pool.

Listing 4.15   Source code for `link_fat12()` ***function.***

```
UCOUNT
link_fat12 (struct dpb *dpbp, UCOUNT Cluster1, UCOUNT Cluster2)
{
    REG UBYTE FAR *fbp0, FAR *fbp1;
    UCOUNT idx;
    struct buffer FAR *bp, FAR *bp1;

    /* Get the block that this cluster is in              */
    bp = getblock((LONG)(((((Cluster1 << 1) + Cluster1) >> 1) /
                    dpbp -> dpb_secsize + dpbp -> dpb_fatstrt
                    + 1), dpbp -> dpb_unit);
    if(bp == NULL)
        return DE_BLKINVLD;

    /* form an index so that we can read the block as a   */
    /* byte array                                         */
    idx = ((((Cluster1 << 1) + Cluster1) >> 1) % dpbp ->
dpb_secsize;

    /* Test to see if the cluster straddles the block. If it  */
    /* does, get the next block and use both to form the      */
    /* FAT word. Otherwise, just point to the next block.     */
    if(idx >= dpbp -> dpb_secsize - 1)
    {
        bp1 = getblock((LONG)(dpbp -> dpb_fatstrt + (((((Cluster1 <<
1)
                    + Cluster1) >> 1) / dpbp -> dpb_secsize))
                    + 2, dpbp -> dpb_unit);
        if(bp1 == (struct buffer *)0)
            return DE_BLKINVLD;
        bp1 -> b_update = TRUE;
        fbp1 = (UBYTE FAR *)&(bp1 -> b_buffer[0]);
    }
    else
        fbp1 = (UBYTE FAR *)&(bp -> b_buffer[idx + 1]);
    fbp0 = (UBYTE FAR *)&(bp -> b_buffer[idx]);
    bp -> b_update = TRUE;
```

Like its 16-bit counterpart, the link_fat12() (Listing 4.15) function begins by getting the block that corresponds to the given FAT number. It also determines the block and fetches it. Next, it computes an index into the block, but unlike link_fat16(), it must pack the desired cluster into the block. To do this, link_fat12() uses an algorithm similar to that used by next_cl12(). Once it packs the cluster number into the block, it must mark the block and update the dpb in the same way as link_fat16().

Listing 4.15   Source code for link_fat12() function —
continued.

```
/* Now pack the value in */
if (Cluster1 & 0x01)
{
    *fbp0 = (*fbp0 & 0x0f) | ((Cluster2 & 0x0f) << 4);
    *fbp1 = (Cluster2 >> 4) & 0xff;
}
else
{
    *fbp0 = Cluster2 & 0xff;
    *fbp1 = (*fbp1 & 0xf0) |(Cluster2 >> 8) & 0x0f;
}

/* update the free space count */
if(Cluster2 == FREE)
{
    /* update the free space count for returned cluster    */
    if(dpbp -> dpb_nfreeclst != UNKNCLUSTER)
        ++dpbp -> dpb_nfreeclst;
}
else
{
    /* update the free space count for removed cluster    */
    if(dpbp -> dpb_nfreeclst != UNKNCLUSTER)
        --dpbp -> dpb_nfreeclst;
}

return SUCCESS;
}
```

Directory Management

Now that I have examined the functions that manage the disk FAT, I will move on to the directory management functions. This really puts you in an excellent position because you have seen the interaction of functions at this level with the getblock() support function. Many of the other functions at this level will operate in a similar fashion. You will also get a first look at a DOS service function. These functions typically have a DOS-like functionality associated with them but may deal with an internal data structure or may only supply a portion of the DOS function.

The dirent *Structure*

Two important structures are used in directory management: dirent and f_node (Listings 4.16 and 4.17). The dirent structure is the first structure I will examine, and it is also a key data structure. The structure's design mirrors the FAT directory entry so that you can easily transfer directory data into and out of memory. It contains filename and extension in fields dir_name and dir_ext identical to the directory record on the disk. DOS-C maintains attributes, date, and time stamp in fields dir_attrib, dir_date, and dir_time that are in memory identical to

---

Listing 4.16   Source code for dirent **structure.**

```
struct dirent
{
    UBYTE                         /* Filename                  */
          dir_name[FNAME_SIZE];
    UBYTE                         /* Filename extension        */
          dir_ext[FEXT_SIZE];
    UBYTE                         /* File Attribute            */
          dir_attrib;
    BYTE  dir_reserved[10];       /* reserved                  */
    time  dir_time;               /* Time file created/updated */
    date  dir_date;               /* Date file created/updated */
    UWORD dir_start;              /* Starting cluster          */
                                  /* 1st available = 2         */
    ULONG dir_size;               /* File size in bytes        */
};
```

their respective directory record entries. However DOS-C translates file size and starting cluster in `dir_size` and `dir_start` to the machine's native memory representation. For 80x86 machines, the memory format matches the disk format, but the code uses portable techniques for other architectures. As you will see, DOS-C uses this structure for every file access and maintenance routine.

Listing 4.17   Source code for `f_node` structure.

```
struct f_node
{
    UWORD f_count;        /* number of uses of this file  */
    COUNT f_mode;         /* read, write, read-write, etc  */
    struct
    {
        BOOL              /* directory has been modified  */
            f_dmod:1;
        BOOL              /* directory is the root        */
            f_droot:1;
        BOOL              /* f_node is new and needs fill  */
            f_dnew:1;
        BOOL              /* f_node is assigned to dir    */
            f_ddir:1;
        BOOL              /* directory is full            */
            f_dfull:1;
    }   f_flags;          /* file flags                   */
    struct dirent         /* this file's dir entry image  */
        f_dir;
    ULONG f_diroff;       /* offset of the dir entry      */
    UWORD f_dirstart;     /* the starting cluster of dir  */
                          /* when dir is not root         */
    struct dpb *          /* the block device for file    */
        f_dpb;
    ULONG f_dsize;        /* file size (for directories)  */
    ULONG f_offset;       /* byte offset for next op      */
    ULONG f_highwater;    /* the largest offset ever      */
    UWORD f_back;         /* the cluster we were at       */
    UWORD f_cluster;      /* the cluster we are at        */
    UWORD f_relcluster;   /* the relative cluster         */
    UWORD f_sector;       /* the sector in the cluster    */
    UWORD f_boff;         /* the byte in the cluster      */
};
```

The f_node *Structure*

The f_node structure is the central data structure that every file and directory operation in fs uses. Within fs, there is a one-to-one relationship between the f_node (file node) structure and any file or directory. Understanding this structure is key to understanding fs itself.

The first member in the structure is f_count. This member maintains a count of the number of references (open, create, etc.) made to this file. When this member is zero, the f_node is free. The next member is f_mode. This member contains a flag that indicates what mode (read, write, read-write, etc.) the file is in.

Many operations within fs go through states. For example, a file must be opened before it can be read. The state that it is in when open must be carried through from one operation to another. Additionally, other more subtle states need storage. The member f_flags contains these bits. The first bit of importance is f_dmod, which indicates that the directory has been modified. This bit is examined when DOS-C closes a file and the directory entry is updated if it is set. The next bit is f_droot, which indicates that the directory associated with this f_node is the root directory and requires special handling. When an f_node is first allocated, a bit is necessary to indicate that the f_node is new and needs to be filled. Bit f_dnew indicates this state. Also, an f_node can be assigned either to a file or a directory. If bit f_ddir is TRUE, the f_node is assigned to a directory. Finally, on occasion, a directory search may encounter an "end of directory" condition, a state especially important when trying to obtain a free directory slot. Bit f_dfull indicates that the directory is full when TRUE.

The next member in the structure is the dirent f_dir structure. This is the file's directory entry image. Every file operation, whether directory management or data manipulation, uses this entry. Associated with the directory image are two members that position the entry within its directory: f_diroff and f_dirstart. The member f_diroff records the byte offset of the directory entry. It is used primarily for locating the correct block and offset for directory read and write operations. Member f_dirstart records the starting cluster of the directory when it is not the disk's root directory. As discussed in Chapter 2, the root directory is a sequential set of disk blocks fixed in both position and length, whereas subdirectories are special files that obey all file rules. As a result, functions that read and write the contents of a directory must be aware of the distinction. The member f_dirstart is an optimization entry that allows DOS-C to do the disk search to find the start only once. As you will see later, this search can be costly for files that are many directories deep.

The next entry is a pointer to a dpb struct. This is the disk parameter block for the block device corresponding to this file. dpb is a data structure that contains all disk-related information such as disk size, geometry, etc. It is this pointer that DOS-C uses to identify the disk. In fact, this is the structure built up by the Media Check Device Driver call.

The next member, f_dsize, is a measure of the size in byes of directories. DOS-C uses this member to maintain the size of the directory during directory searches. It sets this member when opening a directory and checks it when reading a directory. DOS-C also maintains this f_node member in Find First and Find Next functions, because these functions use the Directory Read function in their search.

The next two members in the f_node data structure maintain the file position information. The first member, f_offset, records the byte offset in the file for the next read or write operation. At the beginning of each data transfer operation, DOS-C translates f_offset into information on the physical position of the block and computes the index into the block. In addition, with each write operation, DOS-C compares f_offset to f_highwater and transfers f_offset to f_highwater if f_offset is greater than f_highwater. In this way f_highwater is the write high-water mark, recording the largest offset ever encountered in f_offset. You need to record this information in the event that an application performs a seek operation prior to a file write operation. If DOS-C writes data to the file in a way that extends the file length, and a seek operation follows that moves f_offset to a position within the file, f_highwater contains the true file size. DOS-C later uses f_highwater to update the directory entry when the application closes the file.

The member f_back contains the last cluster number of a sequential file access. It is used to link the FAT when new blocks are added to the file. Associated with this member is f_cluster. This member contains the current cluster number and is used to map the current position of the next read or write operation performed on the file.

In order to map the file offset into the physical sector and the byte offset into the sector, structure members f_sector and f_boff are used to map the sector into the cluster and the byte offset into the cluster, respectively. These members are necessary because they must be maintained between read and write operations in order to lessen the overhead for each operation.

Opening a Directory

Now that I have examined the fundamental f_node data structure, I can begin to look at the file system functions. These functions are the first of the file system functions I will cover that make use of the f_node structure. They are also similar to the FAT functions in that they manage the second part of a FAT file.

DOS-C treats directories in a fashion similar to files. A directory is referenced identically to a file. For example, a directory has a path name that references the directory. It also has a directory entry in a directory with one notable exception — the root directory. Its content is a sequential set of records that are randomly accessed and updated as needed. In short, a directory resembles a file in many respects. For all these reasons, DOS-C manages directories in a fashion similar to files.

Opening a file entails allocating an f_node kernel data structure and initializing it through a call to dir_open() (Listing 4.18). dir_open() does this by first allocating an f_node structure and presetting it to read-write mode. Any f_node that represents a directory must be placed into read-write mode so that DOS-C properly records any update made to the file by subsequent operations.

Next you determine what drive you are using so that you can index the block device table. You need this information because this data structure contains the entire state of the block device which may be modified by the user (e.g., change the current directory, write to a file, etc.).

Listing 4.18   Source code for `dir_open()` ***function.***

```
struct f_node *
dir_open (BYTE FAR *dirname)
{
    struct f_node *fnp;
    WORD drive;
    REG struct dpb *dpbp;
    struct f_node *get_f_node();
    BYTE path[64], *s;
    /* Allocate an f_node if possible - error return (0) if not.*/
    if((fnp = get_f_node()) == (struct f_node *)0)
    {
        return (struct f_node *)NULL;
    }

    /* Force the f_node into read-write mode                    */
    fnp -> f_mode = RDWR;

    /* Determine what drive we are using...                     */

    /* for FAT-style file systems, if the second character of   */
    /* of the path == ':', then a drive specifcation was        */
    /* issued. If it was, update the path to point past the     */
    /* file specification and assign drive to the requested     */
    /* drive, otherwise ...                                     */
    if(*(dirname + 1) == ':')
    {
        drive = *dirname - 'A';
        if(drive > 26)
            drive -= ('a' - 'A');
        dirname += 2;
        dpbp = &blk_devices[drive];
    }
    else
    {
    /* Select the default to help non-drive specified path     */
    /* searches...                                             */
        dpbp = &blk_devices[drive = default_drive];
    }
    fnp -> f_dpb = dpbp;
    /* Generate full path name                                 */
    if(*dirname == '/' || *dirname == '\\')
        fscopy(dirname, (BYTE FAR *)path);
```

You then generate the full path name so that you can properly locate the directory that contains its entry. Remember that a user can specify an absolute filename (i.e., `c:\dos\format.exe`) or a relative filename (i.e., `..\my_dir\my_file.txt`). You parse the filename prepending the current directory for that drive if necessary, then clean up the path by eliminating redundant path separators and relative directory names such as ".." and "..".

Listing 4.18 Source code for `dir_open()` function — continued.

```
else
{
    /* start with the logged in directory                */
    scopy(dpbp -> dpb_path, path);
    /* and append passed relative directory               */
    for(s = path; *s != '\0'; ++s);
    *s++ = '\\';
    fscopy(dirname, (BYTE FAR *)s);
}
/* then clean up the path                                 */
trim_path(path);
/* Determine if we are starting from the root...          */

/* For FAT-style file systems, the root is a consective   */
/* number of blocks given by the bpb. All sub-directories  */
/* are files and need to be treated as such.               */
fnp -> f_flags.f_droot = (*path == '/' || *path == '\\')
                    && (*(path + 1) == '\0');

/* Perform all directory common handling after all         */
/* special handling has been performed.                    */

++dpbp -> dpb_count;
if(media_check(dpbp) < 0)
{
    --dpbp -> dpb_count;
    release_f_node(fnp);
    return (struct f_node *)0;
}
```

Now the fun begins. You need to determine if you are starting from the root because for FAT-style file systems, the root is a consecutive number of blocks given by the dpb. However, all subdirectories are files and dir_open() handles them as such. Because of the differences

---

Listing 4.18   Source code for dir_open() function —
continued.

```
fnp -> f_diroff = 01;
fnp -> f_flags.f_dmod = FALSE;    /* a brand new f_node      */
fnp -> f_flags.f_dnew = TRUE;
fnp -> f_dsize = DIRENT_SIZE * dpbp -> dpb_dirents;
if(!fnp -> f_flags.f_droot)
{
    BYTE dbuff[FNAME_SIZE+FEXT_SIZE], *p;
    WORD i;
    /* Walk the directory tree to find the starting cluster */
    /*                                                       */
    /* Set the root flags since we always start             */
    /* from the root                                        */
    fnp -> f_flags.f_droot = TRUE;
    for(p = path; *p != '\0'; )
    {
        /* skip all path seperators                         */
        while(*p == '/' || *p == '\\')
            ++p;
        /* don't continue if we're at the end               */
        if(*p == '\0')
            break;

        /* Convert the name into an absolute                */
        /* name for comparison...                           */
        /* first the file name with trailing spaces...      */
        for(i = 0; i < FNAME_SIZE; i++)
        {
            if(*p != '\0' && *p != '.' && *p != '/' && *p != '\\')
                dbuff[i] = *p++;
            else
                break;
        }
```

between a FAT root directory and subdirectory, the initial task of finding the starting point of the directory differs, and `dir_open()` handles each case individually.

**Listing 4.18   Source code for `dir_open()` function —
continued.**

```
for( ; i < FNAME_SIZE; i++)
    dbuff[i] = ' ';

/* and the extension (don't forget to            */
/* add trailing spaces)...                        */
if(*p == '.')
    ++p;
for(i = 0; i < FEXT_SIZE; i++)
{
    if(*p != '\0' && *p != '.' && *p != '/' && *p != '\\')
        dbuff[i+FNAME_SIZE] = *p++;
    else
        break;
}
for( ; i < FEXT_SIZE; i++)
    dbuff[i+FNAME_SIZE] = ' ';

/* Now search through the directory to            */
/* find the entry...                              */
i = FALSE;
touc((BYTE FAR *)dbuff, FNAME_SIZE+FEXT_SIZE);
while(dir_read(fnp) == DIRENT_SIZE)
{
    if(fnp -> f_dir.dir_name[0] != '\0' && fnp ->
        f_dir.dir_name[0] != DELETED)
    {
        if(fcmp((BYTE FAR *)dbuff, (BYTE FAR *)fnp ->
                f_dir.dir_name, FNAME_SIZE+FEXT_SIZE))
        {
            i = TRUE;
            break;
        }
    }
}
```

After `dir_open()` completes all special handling it performs all common directory handling. It increments the reference count for the dpb in order to keep track of the number of files using this block device. It also performs a media check to make sure the user did not change the media in the drive, and it corrects internal data structures if they are different. This is one important difference between MS-DOS-style operating systems and UNIX-style operating systems. MS-DOS-style operating systems allow you to change the disk on the fly, whereas

---

Listing 4.18   Source code for `dir_open()` function — continued.

```
            if(!i || !(fnp -> f_dir.dir_attrib & D_DIR))
            {
                --dpbp -> dpb_count;
                release_f_node(fnp);
                return (struct f_node *)0;
            }
            else
            {
                /* make certain we've moved off root       */
                fnp -> f_flags.f_droot = FALSE;
                fnp -> f_flags.f_ddir = TRUE;
                /* set up for file read/write              */
                fnp -> f_offset = 0l;
                fnp -> f_highwater = 0l;
                fnp -> f_cluster = fnp -> f_dir.dir_start;
                fnp -> f_dirstart = fnp -> f_dir.dir_start;
                /* reset the directory flags               */
                fnp -> f_diroff = 0l;
                fnp -> f_flags.f_dmod = FALSE;
                fnp -> f_flags.f_dnew = TRUE;
                fnp -> f_dsize = DIRENT_SIZE * dpbp -> dpb_dirents;
            }
        }
    }
    return fnp;
}
```

UNIX-style operating systems require that the disk be mounted prior to use. Any file system open or create will always do a media check as a pseudo-mount operation.

If the media check is successful, dir_open() initializes the f_node data structures to point to the beginning of the directory, ready for a read operation. Part of this initialization is walking the directory tree to find the starting cluster. Remember that the search may need to go through several directories, verifying the path, in order to find the starting cluster. This search first sets the root flags because the trimmed path is an absolute path that always starts from the root directory. dir_open() adjusts each component of the path to match an MS-DOS directory name entry by adding trailing spaces and concatenating the extension. It then performs a comparison between the adjusted name and the name currently in the f_node. When a match is found, dir_open() records the starting cluster into the f_node. dir_open() repeats this process until the path is exhausted and the f_node contains the starting cluster of the full path.

Listing 4.19 Source code for dir_read() and dir_write() functions.

```
COUNT
dir_read (REG struct f_node *fnp)
{
    REG i, j;
    struct buffer FAR *bp;
    /* Directories need to point to their current offset,    */
    /* not for next op. Therefore, if it is anything other   */
    /* than the first directory entry, we will update the    */
    /* offset on entry rather than wait until exit. If it    */
    /* was new, clear the special new flag.                  */
    if(fnp -> f_flags.f_dnew)
        fnp -> f_flags.f_dnew = FALSE;
    else
        fnp -> f_diroff += DIRENT_SIZE;
```

When dir_open() successfully completes, it returns a pointer to the initialized f_node data structure. This is an important feature to note because this pointer is the internal representation of a file. All file operations work on this internal data structure and DOS-C converts all user handles to an f_node pointer through the interface functions for internal file operations.

Reading/Writing a Directory Entry

Once a directory is opened, it can either be read from or written to. DOS-C provides two functions for this purpose: dir_read() and dir_write() (Listing 4.19). DOS-C only uses these functions in file

---

Listing 4.19   Source code for *dir_read()* and *dir_write()* functions — continued.

```
/* Determine if we hit the end of the directory. If we    */
/* have, bump the offset back to the end and exit. If      */
/* not, fill the dirent portion of the f_node, clear the   */
/* f_dmod bit and leave, but only for root directories     */
if(!(fnp -> f_flags.f_droot)
   && fnp -> f_diroff >= fnp -> f_dsize)
{
    fnp -> f_diroff -= DIRENT_SIZE;
    return 0;
}
else
{
    if(fnp -> f_flags.f_droot)
    {
        if((fnp -> f_diroff / fnp -> f_dpb -> dpb_secsize
            + fnp -> f_dpb -> dpb_dirstrt)
            >= fnp -> f_dpb -> dpb_data)
        {
            fnp -> f_flags.f_dfull = TRUE;
            return 0;
        }
        bp = getblock((LONG)(fnp -> f_diroff / fnp -> f_dpb ->
                    dpb_secsize + fnp -> f_dpb ->
dpb_dirstrt),
                    fnp -> f_dpb -> dpb_unit);
    }
```

Listing 4.19   Source code for `dir_read()` **and** `dir_write()`
            functions — continued.

```
else
{
    REG UWORD secsize = fnp -> f_dpb -> dpb_secsize;
    /* Do a "seek" to the directory position           */
    fnp -> f_offset = fnp -> f_diroff;
    /* Search through the FAT to find the block         */
    /* that this entry is in.                           */
    if(map_cluster(fnp, XFR_READ) != SUCCESS)
    {
        fnp -> f_flags.f_dfull = TRUE;
        return 0;
    }

    /* If the returned cluster is FREE, return zero     */
    /* bytes read.                                      */
    if(fnp -> f_cluster == FREE)
        return 0;

    /* If the returned cluster is LAST_CLUSTER or       */
    /* LONG_LAST_CLUSTER, return zero bytes read        */
    /* and set the directory as full.                   */
    if(last_link(fnp))
    {
        fnp -> f_diroff -= DIRENT_SIZE;
        fnp -> f_flags.f_dfull = TRUE;
        return 0;
    }
    /* Compute the block within the cluster and the     */
    /* offset within the block.                         */
    fnp -> f_sector =
     (fnp -> f_offset / secsize)
      & fnp -> f_dpb -> dpb_clsmask;
    fnp -> f_boff = fnp -> f_offset % secsize;

    /* Get the block we need from cache                 */
    bp = getblock(
        (LONG)clus2phys(fnp -> f_cluster,
            fnp -> f_dpb -> dpb_clssize,
            fnp -> f_dpb -> dpb_data)
            + fnp -> f_sector,
        fnp -> f_dpb -> dpb_unit);
}
```

open, create, and close operations and they cannot be accessed by the user. They are only called in an indirect fashion. As you have done before, start with the read function and then note the symmetry and differences between the read and write functions.

Before I dive into the operation, you should note one difference between directory operations and normal file operations. Data read from or written to a directory is contained within the f_node data structure contained in kernel memory, not in user memory. As a result, DOS-C directories need to point to their current offset because of the record nature of the directory entry. This means that if it is anything other than the first directory entry, you update the offset on entry then wait until exit as you would do for normal file data operations.

Listing 4.19   Source code for `dir_read()` **and** `dir_write()` **functions — continued.**

```
        /* Now that we have the block for our entry, get      */
        /* the directory entry.                               */
        if(bp != NULL)
            getdirent((BYTE FAR *)&bp -> b_buffer[fnp ->
                    f_diroff % fnp -> f_dpb -> dpb_secsize],
                    (struct dirent FAR *)&fnp -> f_dir);
        else
        {
            fnp -> f_flags.f_dfull = TRUE;
            return 0;
        }

        /* Update the f_node's directory info                 */
        fnp -> f_flags.f_dfull = FALSE;
        fnp -> f_flags.f_dmod = FALSE;

        /* and for efficiency, stop when we hit the first     */
        /* unused entry.                                      */
        if(fnp -> f_dir.dir_name[0] == '\0')
            return 0;
        else
            return DIRENT_SIZE;
    }
}
```

Following this rule, dir_read() starts exactly this way. The algorithm also must determine if you hit the end of the directory. If you have, you need to make certain that you do not proceed past the last record; otherwise, you fill the dirent portion of the f_node for root directories or do a "seek" to the directory position and read the cluster number. If the returned cluster is FREE or it is LAST_CLUSTER, you return zero bytes read; otherwise, you compute the block within the cluster and offset within the block. You then read the directory entry into the dirent portion of the f_node and return DIRENT_SIZE bytes read.

Function dir_write() functions slightly differently. It begins by determining if the f_node was modified by a write or create operation. If so, the function updates the disk block containing the entry and marks the buffer so that the buffer cache functions will write the block to disk when needed. This is simple for root directories because they are sequential disk blocks and a seek is a simple computation. All other directories require different handling.

Listing 4.19   Source code for dir_read() **and** dir_write()
              functions — continued.

```
COUNT
dir_write (REG struct f_node *fnp)
{
    struct buffer FAR *bp;

    /* Update the entry if it was modified by a write or create... */
    if(fnp -> f_flags.f_dmod)
    {
        /* Root is a consecutive set of blocks, so handling    */
        /* is simple.                                          */
        if(fnp -> f_flags.f_droot)
        {
            bp =
            getblock(
            (LONG)(fnp -> f_diroff / fnp -> f_dpb -> dpb_secsize
                + fnp -> f_dpb -> dpb_dirstrt),
                fnp -> f_dpb -> dpb_unit);
        }
```

Listing 4.19   *Source code for* `dir_read()` *and* `dir_write()`
**   *functions — continued.***

```
        /* All other directories are just files. The only      */
        /* special handling is resetting the offset so that     */
        /* we can continually update the same directory entry.  */
        else
        {
            REG UWORD secsize = fnp -> f_dpb -> dpb_secsize;

            /* Do a "seek" to the directory position            */
            /* and convert the f_node to a directory f_node.    */
            fnp -> f_offset = fnp -> f_diroff;
            fnp -> f_back = LONG_LAST_CLUSTER;
            fnp -> f_cluster = fnp -> f_dirstart;
            /* Search through the FAT to find the block          */
            /* that this entry is in.                            */
            if(map_cluster(fnp, XFR_READ) != SUCCESS)
            {
                fnp -> f_flags.f_dfull = TRUE;
                release_f_node(fnp);
                return 0;
            }
            /* If the returned cluster is FREE, return zero      */
            /* bytes read.                                       */
            if(fnp -> f_cluster == FREE)
            {
                release_f_node(fnp);
                return 0;
            }

            /* Compute the block within the cluster and the      */
            /* offset within the block.                          */
            fnp -> f_sector =
             (fnp -> f_offset / secsize)
               & fnp -> f_dpb -> dpb_clsmask;
            fnp -> f_boff = fnp -> f_offset % secsize;

            /* Get the block we need from cache                  */
            bp = getblock(
                (LONG)clus2phys(fnp -> f_cluster,
                    fnp -> f_dpb -> dpb_clssize,
                    fnp -> f_dpb -> dpb_data)
                    + fnp -> f_sector,
                fnp -> f_dpb -> dpb_unit);
        }
```

Note that all directories other than root are just files. However, you do need to remember that directory operations are record-style operations. This means that the only special handling is resetting the offset so that you can continually update the same directory entry. dir_write() handles the seek to the correct record in a similar fashion to the dir_read() algorithm, and the disk block is updated. When you successfully complete, you return DIRENT_SIZE bytes written.

Closing a Directory

dir_close() is the function responsible for closing a directory (Listing 4.20). Because of the f_node design, it is surprisingly simple. First, dir_close() tests for invalid f_nodes to guarantee code integrity, although the f_node should be valid at this point. If the f_node contains garbage or is an invalid pointer, a kernel crash or corruption of the file system could be the result, so dir_close() guards against invalid f_nodes.

Listing 4.19   Source code for dir_read() and dir_write() functions — continued.

```
        /* Now that we have a block, transfer the diectory   */
        /* entry into the block.                             */
        if(bp == NULL)
        {
            release_f_node(fnp);
            return 0;
        }
        putdirent((struct dirent FAR *)&fnp -> f_dir,
                (VOID FAR *)&bp -> b_buffer[fnp ->
                f_diroff % fnp -> f_dpb -> dpb_secsize]);
        bp -> b_update = TRUE;
    }
    return DIRENT_SIZE;
}
```

dir_close() then proceeds to write the entry through a call to dir_write(). This is a conditional write because the dir_write() function examines the f_node to see if a prior call modified it and only performs the write if necessary. This optimization helps performance by reducing unnecessary disk access. Remember, disks are electromechanical devices and their speed is very slow when compared to processor speeds.

Next, dir_close() updates the buffer cache by calling flush_buffers(), causing all modified disk buffers, which include both file and directory data, to be written to disk. Again, flush_buffers() only conditionally updates modified buffers for optimization purposes. Once dirty buffers in the buffer cache are safely on disk, dir_close() decrements the reference count and releases the instance of the f_node. Note that in order to correctly share files when users spawn child processes, release_f_node() only returns the f_node to the f_node pool when the reference count is zero.

Listing 4.20   Source code for dir_close() **function.**

```
VOID
dir_close (REG struct f_node *fnp)
{
    REG COUNT disk = fnp -> f_dpb -> dpb_unit;

    /* Test for invalid f_nodes                    */
    if(fnp == NULL)
        return;

#ifndef IPL
    /* Write out the entry                         */
    dir_write(fnp);
#endif

    /* Clear buffers after release                 */
    flush_buffers(disk);
    setinvld(disk);

    /* and release this instance of the f_node     */
    --(fnp -> f_dpb) -> dpb_count;
    release_f_node(fnp);
}
```

File Management

Find First/Next

I now look at the first DOS support functions. Unlike UNIX-like operating systems, a user cannot directly access an MS-DOS directory. However, the need for reading the directory contents is important. A user may want to look for a specific file, get the date or access rights of a file, or simply list the contents of the directory. The designers of MS-DOS designed the dos_findfirst() and dos_findnext() functions to handle these situations (Listing 4.21).

Listing 4.21   Source code for `dos_findfirst()` ***and***
               `dos_findnext()` ***functions.***

```
COUNT
dos_findfirst (UCOUNT attr, BYTE FAR *name)
{
    REG struct f_node *fnp;
    REG dmatch FAR *dmp = (dmatch FAR *)dta;
    struct dosnames DosName;
    REG COUNT i;
    BYTE *p;

    /* The findfirst/findnext calls are probably the worst   */
    /* of the DOS calls. They must work somewhat on the fly  */
    /* (i.e. - open but never close). Since we don't want to  */
    /* lose f_nodes every time a directory is searched, we will */
    /* initialize the DOS dirmatch structure and then for    */
    /* every find, we will open the current directory, do a  */
    /* seek and read, then close the f_node.                 */

    /* Start out by initializing the dirmatch structure.     */
    dmp -> dm_drive = default_drive;
    dmp -> dm_entry = 0;
    dmp -> dm_cluster = 0;

    dmp -> dm_attr_srch = attr;

    if(DosNames(name, (struct dosnames FAR *)&DosName) != SUCCESS)
        return DE_FILENOTFND;
```

These function calls must work somewhat on the fly. If you do not properly design the algorithm, dos_findfirst() and dos_findnext() could end up opening a directory and reading it but never closing it. This would cause you to loose an f_node for every directory search. To open a directory and never close it, you must properly implement the support functions, dir_open(), dir_read(), and dir_write().

Listing 4.21   Source code for dos_findfirst() and dos_findnext() functions — continued.

```
dmp -> dm_drive = DosName.dn_drive;

/* Build the match pattern out of the passed string        */
for(p = DosName.dn_name, i = 0; i < FNAME_SIZE; i++)
{
    /* test for a valid file name terminator               */
    if(*p != '\0' && *p != '.')
    {
        /* If not a wildcard ('*'), just transfer          */
        if(*p != '*')
            dmp -> dm_name_pat[i] = *p++;
        else
        {
            /* swallow the wildcard                         */
            ++p;

            /* fill with character wildcard (?)             */
            for( ; i < FNAME_SIZE; i++)
                dmp -> dm_name_pat[i] = '?';
            /* and skip to seperator                        */
            while(*p != '\0' && *p != '.' && *p
                != '/' && *p != '\\')
                ++p;

            break;
        }
    }
    else
        break;
}
```

Because you don't want to lose f_nodes every time an application searches a directory, you initialize the MS-DOS dirmatch structure and then for every find, you open the current directory, do a seek and read, then close the f_node. You also store all necessary information in the dirmatch structure that is in user space. This way, the user indirectly handles f_node resource management.

Look a little closer at dos_findfirst(). The function begins by initializing the dirmatch structure. Next, it parses the pattern in order to create an expanded pattern in the dirmatch structure to use in the

---

**Listing 4.21   Source code for dos_findfirst() and
dos_findnext() functions — continued.**

```
for( ; i < FNAME_SIZE; i++)
    dmp -> dm_name_pat[i] = ' ';

i = 0;

/* and the extension (don't forget to add trailing spaces)... */
if(*p == '.')
{
    ++p;
    for( ; i < FEXT_SIZE; i++)
    {
        if(*p != '\0' && *p != '.' && *p != '/' && *p != '\\')
        {
            if(*p != '*')
                dmp -> dm_name_pat[i+FNAME_SIZE] = *p++;
            else
            {
                for( ; i < FEXT_SIZE; i++)
                    dmp -> dm_name_pat[i+FNAME_SIZE] = '?';
                break;
            }
        }
        else
            break;
    }
}
```

directory reads. Note that the expanded pattern removes all "*" characters and fills the pattern with the character wildcard "?". The algorithm later used in dos_find() next automatically matches the character wildcard in the source pattern. This is how DOS-C implements wildcards.

Listing 4.21   Source code for* dos_findfirst() *and
dos_findnext() *functions — continued.

```
for( ; i < FEXT_SIZE; i++)
    dmp -> dm_name_pat[i+FNAME_SIZE] = ' ';

/* Now search through the directory to find the entry...  */
touc((BYTE FAR *)dmp -> dm_name_pat, FNAME_SIZE+FEXT_SIZE);

/* Special handling - the volume id is only in the root   */
/* directory and only searched for once.  So we need to   */
/* open the root and return only the first entry that     */
/* contains the volume id bit set.                        */
if(attr & D_VOLID)
{
    /* Now open this directory so that we can read the    */
    /* f_node entry and do a match on it.                 */
    if((fnp = dir_open((BYTE FAR *)"\\")) == NULL)
        return DE_PATHNOTFND;
    /* Now do the search                                  */
    while(dir_read(fnp) == DIRENT_SIZE)
    {
        /* Test the attribute and return first found      */
        if(fnp -> f_dir.dir_attrib & D_VOLID)
        {
            pop_dmp(dmp, fnp);
            dir_close(fnp);
            return SUCCESS;
        }
    }

    /* Now that we've done our failed search, close it and */
    /* return an error.                                    */
    dir_close(fnp);
    return DE_FILENOTFND;
}
```

With the initialization complete, now search through the directory to find the entry. Use dos_findnext() to return the first match, but first you must perform some special handling. The volume ID is only in the root directory and only searched for once. Whenever the attributes passed in from the user require a volume ID search, you need to open the root directory and return only the first entry where the volume ID bit is set. Open the root directory and search by using a while loop that reads the directory entry and terminates when you reach the end of the directory. Within the loop, test the attribute of the directory entry just read and return the first directory entry found (if any). If you do find a match, dos_findfirst() populates the users dirent structure by

---

Listing 4.21   Source code for dos_findfirst() and dos_findnext() functions — continued.

```
    /* Otherwise just do a normal find next                */
    else
    {
        BYTE LocalPath[67];

        /* Build the full path so that we can open it      */
        sprintf(LocalPath, "%c:%s", 'A' + dmp -> dm_drive,
         *DosName.dn_path == '\0' ? "." : DosName.dn_path);

        /* Now open this directory so that we can read the */
        /* f_node entry and do a match on it.              */
        if((fnp = dir_open((BYTE FAR *)LocalPath)) == NULL)
            return DE_PATHNOTFND;

        pop_dmp(dmp, fnp);
        dmp -> dm_entry = 0;
        if(!fnp -> f_flags.f_droot)
            dmp -> dm_cluster = fnp -> f_dirstart;
        else
            dmp -> dm_cluster = 0;
        dir_close(fnp);
        return dos_findnext();
    }
}
```

using the pop_dmp( ) function, closes the file, and returns successfully. If dos_findfirst( ) did not find the entry, it closes the root directory and returns an error.

If the volume ID attribute bit is not set, dos_findfirst( ) uses dos_findnext( ) to find the first match. dos_findfirst( ) prepares for the call by first building the full path so that it can open the directory using dir_open( ). It opens this directory so that it can determine the starting cluster of the directory and save the cluster in the users dirent structure. Finally, it closes the directory, which in turn returns all resources, and exits through dos_findnext( ).

Although dos_findfirst( ) did a search for volume ID, dos_findnext( ) performs the bulk of the work. This partitioning of functionality guarantees identical searches for both functions. This is why dos_findfirst( ) exits through dos_findnext( ).

To start, dos_findnext( ) assigns a pointer to the match parameters. This pointer is actually a pointer to the disk transfer area because MS-DOS specifies that the dirmatch structure be contained within the

---

Listing 4.21   Source code for dos_findfirst( ) and dos_findnext( ) functions — continued.

```
COUNT
dos_findnext (void)
{
    REG dmatch FAR *dmp = (dmatch FAR *)dta;
    REG struct f_node *fnp;
    BOOL found = FALSE;
    BYTE FAR *p, *q;

    /* assign our match parameters pointer.               */
    dmp = (dmatch FAR *)dta;

    /* Allocate an f_node if possible - error return (0) if not.*/
    if((fnp = get_f_node()) == (struct f_node *)0)
    {
        return DE_FILENOTFND;
    }
```

dta (a remnant of the CP/M compatibility rules). Next dos_findnext() allocates an f_node if possible and returns a "File not found" error if it cannot. dos_findnext() next selects the drive parameter block entry that corresponds to the drive specified in the dirmatch structure. dos_findnext() then performs a media check to guarantee that the user did not change the disk between the calls and returns an error if this did happen.

Listing 4.21   Source code for dos_findfirst() and
dos_findnext() functions — continued.

```
/* Force the f_node into read-write mode               */
fnp -> f_mode = RDWR;
/* Select the default to help non-drive specified path */
/* searches...                                         */
fnp -> f_dpb = &blk_devices[dmp -> dm_drive];
++(fnp -> f_dpb) -> dpb_count;

if(media_check(fnp -> f_dpb) < 0)
{
    --(fnp -> f_dpb) -> dpb_count;
    release_f_node(fnp);
    return DE_FILENOTFND;
}

fnp -> f_dsize = DIRENT_SIZE * (fnp -> f_dpb) -> dpb_dirents;

/* Search through the directory to find the entry, but do */
/* a seek first.                                          */
if(dmp -> dm_entry > 0)
    fnp -> f_diroff = (dmp -> dm_entry - 1) * DIRENT_SIZE;

fnp -> f_offset = fnp -> f_highwater = fnp -> f_diroff;

fnp -> f_cluster = dmp -> dm_cluster;
fnp -> f_dirstart = dmp -> dm_cluster;
```

Listing 4.21   Source code for `dos_findfirst()` **and**
                `dos_findnext()` **functions — continued.**

```
/* Loop through the directory                               */
while(dir_read(fnp) == DIRENT_SIZE)
{
    ++dmp -> dm_entry;
    if(fnp -> f_dir.dir_name[0] != '\0' && fnp ->
        f_dir.dir_name[0] != DELETED)
    {
        if(fcmp_wild((BYTE FAR *)(dmp -> dm_name_pat),
                     (BYTE FAR *)fnp -> f_dir.dir_name,
                     FNAME_SIZE+FEXT_SIZE))
        {
            /* Test the attribute as the final step*/
            if(fnp -> f_dir.dir_attrib & D_VOLID)
                continue;
            else if(
              ((~(dmp -> dm_attr_srch | D_ARCHIVE | D_RDONLY)
               & fnp -> f_dir.dir_attrib)
               & (D_DIR | D_SYSTEM | D_HIDDEN)) == 0)
            {
                found = TRUE;
                break;
            }
            else
                continue;
        }
    }
}

/* If found, transfer it to the dmatch structure           */
if(found)
    pop_dmp(dmp, fnp);

/* return the result                                        */
--(fnp -> f_dpb) -> dpb_count;
release_f_node(fnp);

return found ? SUCCESS : DE_FILENOTFND;
}
```

With all these preliminaries out of the way, dos_findnext() gets down to business and starts to search through the directory to find the entry. First it recovers information saved in the dirmatch structure by dos_findfirst() and does a seek to the directory record that corresponds to the next record to be read. dos_findnext() falls into a while loop whose exit criterion is an end-of-file on the directory. As it loops through the directory, it performs a wildcard match on the name field of the entry if it is not a deleted entry. If a match occurs, it tests the attribute as the final step because a match condition requires both name and attribute matches. If an entry meets both criteria, dos_findnext() transfers the entry into the dmatch structure and returns the resources. Finally, it returns a success or file-not-found error based on the result of the search. This satisfies MS-DOS requirements and completes the first look at DOS support functions.

Listing 4.22   Source code for dos_open( ) ***function.***

```
COUNT
dos_open (BYTE FAR *path, COUNT flag)
{
    REG struct f_node *fnp;
    COUNT i;
    BYTE FAR *fnamep;
    BYTE dname[NAMEMAX];
    BYTE fname[FNAME_SIZE], fext[FEXT_SIZE];

    /* First test the flag to see if the user has passed a    */
    /* valid file mode...                                      */
    if(flag < 0 || flag > 2)
        return DE_INVLDACC;

    /* first split the passed dir into comopnents (i.e. - path */
    /* to new directory and name of new directory.             */
    if((fnp = split_path(path, dname, fname, fext)) == NULL)
    {
        dir_close(fnp);
        return DE_PATHNOTFND;
    }
```

Opening a File

If you have followed the directory functions so far, most of what you will see in the file handlers will look very familiar. The DOS-C design reuses code and algorithms wherever possible, facilitated through the use of the f_node data structure common to all the file system functions. I will examine dos_open( ) to study how a file access works (Listing 4.22). The reader can look at other functions such as dos_create( ) because of the similarity in design.

Listing 4.22   Source code for dos_open( ) function —
continued.

```
/* Look for the file. If we can't find it, just return a   */
/* not found error.                                         */
if(!find_fname(fnp, fname, fext))
{
    dir_close(fnp);
    return DE_FILENOTFND;
}

/* Set the f_node to the desired mode                       */
fnp -> f_mode = flag;

/* Initialize the rest of the f_node.                       */
fnp -> f_offset = 0l;
fnp -> f_highwater = fnp -> f_dir.dir_size;

fnp -> f_back = LONG_LAST_CLUSTER;
fnp -> f_cluster = fnp -> f_dir.dir_start;

fnp -> f_flags.f_dmod = FALSE;
fnp -> f_flags.f_dnew = FALSE;
fnp -> f_flags.f_ddir = FALSE;

return xlt_fnp(fnp);
}
```

The algorithm for dos_open() is similar to that of dir_open(). The only changes are that a file also has a mode associated with it and that it must return a number that will be later used to derive the handle returned to the user.

dos_open() begins by first testing the flag to see if the user has passed a valid file mode. This type of testing is necessary for all of the DOS support functions. An application can easily pass erroneous information and DOS-C must handle this in a sane way. Crashing because of bad parameters passed to a call is definitely not an option.

Once parameter testing is complete, the path passed in the call needs to be validated. dos_open() splits the path into components (i.e., path to directory, filename, and file extension) and tests for a valid path. This information is needed to locate the file. If the file can't be found, dos_open() returns a file-not-found error. If the file is found, dos_open() proceeds to set the f_node to the desired mode and initializes the rest of the f_node. It finally returns through xlt_fnp(), which translates an f_node pointer to a numeric index for later use.

With f_node initialized, any file operation can take place, and f_node retains the state of the file. It is this data structure that contains the directory image, current size, pointer to the next byte to be read, etc.

Reading/Writing a File

Reading and writing a file is a large part of the DOS-C file system code. This code is used for loading programs as well as file access to user applications. DOS-C provides two functions, dos_read() and dos_write() for DOS support functions, but a single function, rdwrblock(), does all the work (Listing 4.23). Both dos_read() and dos_write() call rdwrblock() with XFR_READ to read a file and XFR_WRITE to write a file.

rdwrblock() is common to both functions because of the similarity between file read and write operations. In both cases, data is transferred with only data direction differences. The block cache mechanism simplifies matters by providing automatic mechanisms for writing dirty buffers. Hence, data is transferred either from memory to a buffer or vice versa. I will note other small differences along the way.

As you have seen before, rdwrblock() starts off by translating the internal handle into an f_node pointer because all internal file system operations use f_nodes to maintain the state of the file. Part of the translation requires testing the validity of the file. For example, the application may erroneously close the file and attempt to read from it or

---

Listing 4.23   Source code for rdwrblock() *function.*

```
UCOUNT
rdwrblock (COUNT fd, VOID FAR *buffer, UCOUNT count, COUNT mode,
        COUNT *err)
{
    REG struct f_node *fnp;
    REG struct buffer FAR *bp;
    UCOUNT xfr_cnt = 0, ret_cnt = 0;
    LONG idx;
    WORD secsize;
    UCOUNT to_xfer = count;

#ifdef DEBUG
    if(bDumpRdWrParms)
    {
        printf("rdwrblock: mode = %s\n",
            mode == XFR_WRITE ? "WRITE" : "READ");
        printf(" fd    buffer      count\n --  ------    -----\n");
        printf(" %02d   %04x:%04x   %d\n", fd, (COUNT)FP_SEG(buffer),
            (COUNT)FP_OFF(buffer), count);
    }
#endif
    /* Translate the fd into an f_node pointer, since all      */
    /* internal operations are achieved through f_nodes.       */
    fnp = xlt_fd(fd);

    /* If the fd was invalid because it was out of range or the */
    /* requested file was not open, tell the caller and exit    */
    /* note: an invalid fd is indicated by a 0 return           */
    if(fnp == (struct f_node *)0 || fnp -> f_count <= 0)
    {
        *err = DE_INVLDHNDL;
        return 0;
    }
```

write to it. If the internal handle is invalid because it is out of range or the requested file is not open, rdwrblock() notifies the calling DOS support function of the error and exits. Next, rdwrblock() tests that the requested number of bytes for data transfer is not zero. If the count is zero and the mode is XFR_READ, rdwrblock() just exits because a read with a count of zero is a valid call but does not require any further processing. Additionally, a write with a count of zero is a special case that sets the file length to the current length. Another test is to check for a seek past end-of-file on an XFR_READ operation. This is also a valid read operation that must return 0 bytes transferred. Again, it is easy to test this up front and avoid further processing.

Listing 4.23   Source code for rdwrblock() **function —**
**   continued.**

```
/* Test that we are really about to do a data transfer. If  */
/* the count is zero and the mode is XFR_READ, just exit.    */
/* (Any read with a count of zero is a nop).                 */
/*                                                            */
/* A write (mode is XFR_WRITE) is a special case, it sets     */
/* the file length to the current length (truncates it).      */
/*                                                            */
/* NOTE: doing this up front saves a lot of headaches later. */
if(count == 0)
{
    if(mode == XFR_WRITE)
        fnp -> f_highwater = fnp -> f_offset;
    {
        *err = SUCCESS;
        return 0;
    }
}

/* Another test is to check for a seek past EOF on an        */
/* XFR_READ operation.                                        */
if(mode == XFR_READ
 && !fnp -> f_flags.f_ddir
  && (fnp -> f_offset >= fnp -> f_dir.dir_size))
{
    *err = SUCCESS;
    return 0;
}
```

With the special cases out of the way, you begin testing that the operation is valid. First, rdwrblock() tests that the user call passed a valid mode for this f_node. This consists of testing combinations of read-only and read-write in the case of read operations and write-only and read-write for write operations. If these two combinations do not exist, then an invalid access error is returned.

Next, rdwrblock() adjusts the far pointer from user space to supervisor space with a call to the memory management adjust_far() function.

Listing 4.23   ***Source code for*** rdwrblock() ***function —***
                 continued.

```c
    /* test that we have a valid mode for this f_node        */
    switch(mode)
    {
    case XFR_READ:
        if(fnp -> f_mode != RDONLY && fnp -> f_mode != RDWR)
        {
            *err = DE_INVLDACC;
            return 0;
        }
        break;

#ifndef IPL
    case XFR_WRITE:
        if(fnp -> f_mode != WRONLY && fnp -> f_mode != RDWR)
        {
            *err = DE_INVLDACC;
            return 0;
        }
        break;
#endif
    default:
        *err = DE_INVLDACC;
        return 0;
    }

    /* The variable secsize will be used later.              */
    secsize = fnp -> f_dpb -> dpb_secsize;

    /* Adjust the far pointer from user space tp supervisor space*/
    buffer = adjust_far((VOID FAR *)buffer);
```

Although DOS-C is real mode only, the call is still necessary because of the segment architecture. Adjust the segment and offset so that you have a usable range for the transfer. Once the pointer has been adjusted, proceed to transfer the data. `rdwrblock()` uses a block transfer method so that later versions of DOS-C can use memory management.

**Listing 4.23   Source code for `rdwrblock()` function —
            continued.**

```
    /* Do the data transfer. Use block transfer methods so    */
    /* that we can utilize memory management in more complex   */
    /* DOS-C versions.                                         */
    while(ret_cnt < count)
    {
        /* Position the file to the f_node's pointer position. */
        /* This is done by updating the f_node's cluster, block */
        /* (sector) and byte offset so that read or write      */
        /* becomes a simple data move into or out of the block */
        /* data buffer.                                        */
        if(fnp -> f_offset == 01)
        {
#ifndef IPL
            /* For the write case, a newly created file        */
            /* will have a start cluster of FREE. If we're     */
            /* doing a write and this is the first time        */
            /* through, allocate a new cluster to the file.    */
            if((mode == XFR_WRITE)
             && (fnp -> f_dir.dir_start == FREE))
                if(!first_fat(fnp))
                {
                    dir_close(fnp);
                    *err = DE_HNDLDSKFULL;
                    return ret_cnt;
                }
#endif
            /* complete the common operations of               */
            /* initializing to the starting cluster and        */
            /* setting all offsets to zero.                    */
            fnp -> f_cluster = fnp -> f_dir.dir_start;
            fnp -> f_back = LONG_LAST_CLUSTER;
            fnp -> f_sector = 0;
            fnp -> f_boff = 0;
        }
```

rdwrblock() uses a while loop that exits when the bytes transferred equal the requested amount. This guarantees that the correct
number of bytes is transferred. Next, the loop positions the file to the
pointer position of the f_node. This is done by updating the cluster,
block (sector), and byte offset of the f_node so that read or write
becomes a simple data move into or out of the block data buffer. There

---

**Listing 4.23   Source code for `rdwrblock()` function —
              continued.**

```
    /* The more difficult scenario is the (more common)  */
    /* file offset case. Here, we need to take the f_node's */
    /* offset pointer (f_offset) and translate it into a  */
    /* relative cluster position, cluster block (sector)  */
    /* offset (f_sector) and byte offset (f_boff). Once we */
    /* have this information, we need to translate the    */
    /* relative cluster position into an absolute cluster */
    /* position (f_cluster). This is unfortunate because  */
    /* it requires a linear search through the file's FAT */
    /* entries. It made sense when DOS was originally     */
    /* designed as a simple floppy disk operating system  */
    /* where the FAT was contained in core, but now       */
    /* requires a search through the FAT blocks.          */
    /*                                                    */
    /* The algorithm in this function takes advantage of  */
    /* the blockio block buffering scheme to simplify the */
    /* task.                                              */
    else
        switch(map_cluster(fnp, mode))
        {
        case DE_SEEK:
            dir_close(fnp);
            return ret_cnt;

        default:
            dir_close(fnp);
            *err = DE_HNDLDSKFULL;
            return ret_cnt;

        case SUCCESS:
            break;
        }
```

is a special case that must be examined. For write operations, a newly created file will have a start cluster of FREE. Check for this condition because if this is the case, you must allocate a new cluster to the file. Then rdwrblock( ) completes the common operations of initializing to the starting cluster and setting all offsets to zero.

Listing 4.23   Source code for `rdwrblock( )` ***function — continued.***

```
#ifndef IPL
        /* XFR_WRITE case only - if we're at the end, the next  */
        /* FAT is an EOF marker, so just extend the file length */
        if(mode == XFR_WRITE && last_link(fnp))
            if(!extend(fnp))
            {
                dir_close(fnp);
                *err = DE_HNDLDSKFULL;
                return ret_cnt;
            }
#endif

        /* Compute the block within the cluster and the offset  */
        /* within the block.                                    */
        fnp -> f_sector =
        (fnp -> f_offset / secsize) & fnp -> f_dpb -> dpb_clsmask;
        fnp -> f_boff = fnp -> f_offset % secsize;

#ifdef DSK_DEBUG
    printf("%d links; dir offset %ld, starting at cluster %d\n",
            fnp -> f_count,
            fnp -> f_diroff,
            fnp -> f_cluster);
#endif

        /* Do an EOF test and return whatever was transferred   */
        /* but only for regular files in XFR_READ mode          */
        if((mode == XFR_READ) && !(fnp -> f_flags.f_ddir)
          && (fnp -> f_offset >= fnp -> f_dir.dir_size))
        {
            *err = SUCCESS;
            return ret_cnt;
        }
```

**Listing 4.23   Source code for `rdwrblock()` function —
continued.**

```
        /* Get the block we need from cache*/
        bp = getblock((LONG)clus2phys(fnp -> f_cluster,
                    fnp -> f_dpb -> dpb_clssize,
                    fnp -> f_dpb -> dpb_data) + fnp -> f_sector,
                    fnp -> f_dpb -> dpb_unit);
        if(bp == (struct buffer *)0)
        {
            *err = DE_BLKINVLD;
            return ret_cnt;
        }

        /* transfer a block                                    */
        /* Transfer size as either a full block size, or the   */
        /* requested transfer size, whichever is smaller.      */
        /* Then compare to what is left, since we can transfer  */
        /* a maximum of what is left.                          */
        switch(mode)
        {
        case XFR_READ:
            if(fnp -> f_flags.f_ddir)
                xfr_cnt = min(to_xfer, secsize - fnp -> f_boff);
            else
                xfr_cnt = min(min(to_xfer, secsize - fnp -> f_boff),
                            fnp -> f_dir.dir_size - fnp ->
                            f_offset);
            fbcopy((BYTE FAR *)&bp -> b_buffer[fnp -> f_boff],
                    buffer, xfr_cnt);
            break;

#ifndef IPL
        case XFR_WRITE:
            xfr_cnt = min(to_xfer, secsize - fnp -> f_boff);
            fbcopy(buffer,(BYTE FAR *)&bp -> b_buffer[fnp -> f_boff],
                    xfr_cnt);
            bp -> b_update = TRUE;
            break;
#endif

        default:
            *err =  DE_INVLDACC;
            return ret_cnt;
        }
```

The more common scenario is the common file offset case. Here, you need to take the offset pointer of the f_node (f_offset) and translate it into a relative cluster position, cluster block (sector), offset (f_sector), and byte offset (f_boff). Once you have this information, you need to translate the relative cluster position into an absolute cluster position (f_cluster). This is unfortunate because it requires a linear search

---

Listing 4.23   Source code for `rdwrblock()` function —
continued.

```
        /* update pointers and counters                      */
        ret_cnt += xfr_cnt;
        to_xfer -= xfr_cnt;
        fnp -> f_offset += xfr_cnt;
        buffer = add_far((VOID FAR *)buffer, (ULONG)xfr_cnt);
        if(mode == XFR_WRITE &&
           (fnp -> f_offset > fnp -> f_highwater))
             fnp -> f_highwater = fnp -> f_offset;
    }
    *err = SUCCESS;
    return ret_cnt;
}

COUNT
dos_read (COUNT fd, VOID FAR *buffer, UCOUNT count)
{
    COUNT err, xfr;

    xfr = rdwrblock(fd, buffer, count, XFR_READ, &err);
    return err != SUCCESS ? err : xfr;
}


#ifndef IPL
COUNT
dos_write (COUNT fd, VOID FAR *buffer, UCOUNT count)
{
    COUNT err, xfr;

    xfr = rdwrblock(fd, buffer, count, XFR_WRITE, &err);
    return err != SUCCESS ? err : xfr;
}
#endif
```

through the file's FAT entries. This scenario made sense when DOS was originally designed as a simple floppy disk operating system where the FAT was contained in core, but a search through the FAT blocks is required. The algorithm used by rdwrblock() takes advantage of the blockio block buffering scheme to simplify the task. Because the FAT is contained within a block, the block may remain within the cache if there are enough buffers. This minimizes seek and rotational latencies.

After rdwrblock() determines the physical addresses, it performs another test for a special case. In the write case only, rdwrblock() must check for end-of-file before the data transfer can take place. You need to extend the file by allocating another cluster before you can transfer data to it. For this reason, the test must occur before the transfer. If the file is at the end, the next FAT is an EOF marker and rdwrblock() just extends the file's cluster chain.

For read operations, rdwrblock() does an EOF test for every loop iteration. You need to test for end-of-file because it is a valid termination condition. If you encounter this situation, you return a count of whatever was transferred for regular files in read mode.

If it was not an end-of-file, rdwrblock() proceeds to get the block you need from cache and transfer data to or from the block. Transfer size is either a full block size, or the requested transfer size, whichever is smaller. When this transfer is complete, rdwrblock() compares its count to what is left, since you must transfer what is left. rdwrblock() updates pointers and counters and repeats the loop. It returns the number of bytes transferred when the loop terminates.

Closing a File

Much of the dos_close() algorithm (Listing 4.24) will seem very familiar if you have studied the dir_close() function (Listing 4.20). Like the other file system functions, dos_close() first translates the internal handle into a useful pointer. If the corresponding f_node is invalid because the internal handle is out of range or the requested file is not open, dos_close() simply returns an error. It then looks at the file mode. If the file is not read-only, a write may have taken place.

dos_close() updates the directory size and sets a flag that will cause the directory record to be written. It then calls dir_close() to update the directory, if necessary, and exits.

More Functions

By now, you have looked at file system operations that perform data transfer in one form or another. The file system functions also contain other utilitarian functions that return file sizes, file creation or modification dates, times, etc. The functions are all built from building blocks similar to the data transfer functions and can be easily understood if you have understood the explanations of the data transfer functions. MS-DOS is feature-rich with utility functions; it will be useful to study the code in these files.

Listing 4.24   Source code for dos_close() function.

```
COUNT
dos_close (COUNT fd)
{
    struct f_node *fnp;

    /* Translate the fd into a useful pointer              */
    fnp = xlt_fd(fd);

    /* If the fd was invalid because it was out of range or the */
    /* requested file was not open, tell the caller and exit    */
    /* note: an invalid fd is indicated by a 0 return          */
    if(fnp == (struct f_node *)0 || fnp -> f_count <= 0)
        return DE_INVLDHNDL;
    if(fnp -> f_mode != RDONLY)
    {
        fnp -> f_dir.dir_size = fnp -> f_highwater;
        fnp -> f_flags.f_dmod = TRUE;
    }
    fnp -> f_flags.f_ddir = TRUE;

    dir_close(fnp);
    return SUCCESS;
}
```

## Chapter 5

DOS-C Kernel:
Memory Manager
and Task Manager

In this chapter, I will examine two resource managers: the memory manager and the task manager. Both of these managers serve an important role in the operations of DOS-C (Figure 5.1). Each manager provides system call services that are part of the MS-DOS collection of system services. They also provide key internal services necessary to the other DOS-C services. One note: although the code is written to be portable, portions of these managers, especially the task manager, contain 80x86-specific code.

In DOS-C, the memory manager provides and maintains the memory allocation of DOS-C. Unlike other operating systems, the memory manager does not provide protection to the memory it manages. It cannot, because DOS-C is designed to work in real mode so that it runs on XT systems, which are powered with the 8088 processor. Memory protection mechanisms are not available in the low-end 8088 and in the real

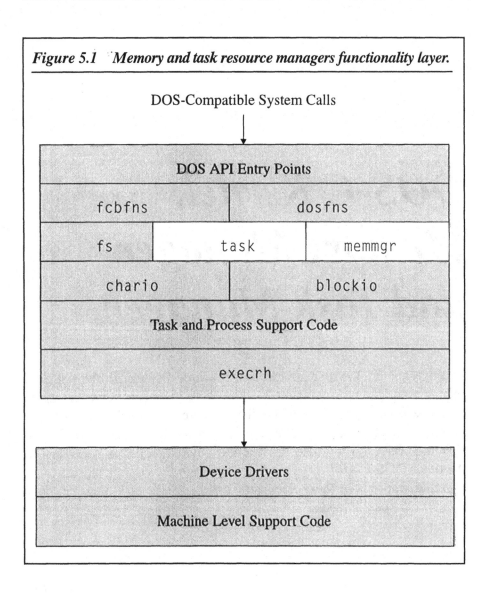

Figure 5.1   Memory and task resource managers functionality layer.

mode of 80286 and newer processors. However, the algorithms used can easily be adapted (and have been in related operating systems) to provide the underlying allocation mechanism for protected-mode versions.

The task manager is also simplified when compared to task managers of other operating systems. MS-DOS is a simple operating system that can only perform a single task at a time. Because DOS-C is patterned after it, DOS-C also provides support for a single task at a time. Its primary responsibility is to act as a task loader and perform a task switch from kernel mode to user mode and vice versa. Again, the code presented here is derived from more sophisticated multitasking code, and the elements necessary for multitasking are contained in this code. However, what is sorely missing is a scheduler, but one can be easily introduced if you are inclined to do so.

Arena Management

In MS-DOS, the memory area is known as the arena. The arena is a collection of one or more areas of memory that contain a data structure followed by available memory. DOS-C uses the same data structures as MS-DOS to maintain the memory arena. This is necessary, because many programs examine the memory header when they want to walk the memory chain to find other programs that may reside in memory. In fact, much of the "undocumented" literature is devoted to explaining the memory header and presents special programs to dump the header. Because the goal of DOS-C is to be compatible with MS-DOS, the memory header is identical.

Like MS-DOS, DOS-C arranges the memory arena as a collection of variable-length blocks. Each block has a Memory Control Block (MCB) defined by the mcb data structure (Listing 5.1). This header starts with a single byte that is ASCII "M" except in the last block which starts with an ASCII "Z". All memory blocks must have an owner, and DOS-C identifies the owner by assigning the segment of the program segment prefix of the owner. This entry is the m_psp member. Note that the use of the segment is a direct 80x86 architecture derivative.

Another 80x86-dependent entry is the size member, m_size. This member contains the size of the memory block in 16-byte paragraphs that match the 80x86 paragraph size. The user gets the memory block starting at the next byte after the MCB header, which falls, coincidentally, on a paragraph boundary. This field is also instrumental in tracing the MCB chain. If you want to determine where the next MCB header will be, take the address of the block and add m_size to it. The next MCB block starts at this address. Of course you must perform all computations in 80x86 segment:offset fashion. Unfortunately, this is not very portable, but portability must be sacrificed for compatibility purposes. However, the code is written in a portable manner and the use of any other granularity such as memory pages can be substituted. For example, an 80386 or 68040 version can use 4Kb pages that correspond to MMU pages.

The last MCB member is m_name. In a departure from standard MS-DOS, DOS-C places the filename of the owner in this field. This allows easy owner identification of the block when programs that walk the MCB chain dump the header contents. DOS-C, unlike MS-DOS, also clears this field when the MCB becomes available.

Another member is m_fill. This odd little unused portion of the MCB header is a fix that brings the size of the header to fit exactly within one paragraph. Again, because of MS-DOS' strong root in the 80x86 world, assembly language arithmetic is smaller and faster if you can ignore the offset and work with the segment only. Microsoft chose to do this and sacrificed 3 bytes to simplify the math. This is a good software engineering tradeoff that results in cleaner code.

Listing 5.1   The mcb data structure.

```
typedef struct
{
    BYTE  m_type;    /* mcb ype - chain or end              */
    UWORD m_psp;         /* owner id via psp segment        */
    UWORD m_size;        /* size of segment in paragraphs   */
    BYTE  m_fill[3];
    BYTE  m_name[8];     /* owner name limited to 8 bytes   */
} mcb;
```

Allocation Strategy Management

DOS-C provides five functions to manage memory. These functions provide allocation, deallocation, size modification, and validation functions. However, not all functions are available to the user. MS-DOS and DOS-C provide system calls to allocate, change size, and return memory to DOS-C. The other functions provide DOS-C with the ability to find a block of memory large enough to load a program and to verify that the arena is intact. Unfortunately, DOS-C needs this functionality because any program could destroy the arena or DOS-C itself because it lacks memory protection schemes.

Memory Allocation

For memory allocation, DOS-C provides the function DosMemAlloc() (Listing 5.2), which can allocate memory using three different criteria: first fit, best fit, and last fit. It can also return the largest memory block available for program loading (Figure 5.2). The algorithm used is straightforward — search each memory block until it encounters the last one or meets an exit criteria. If a block matches the criteria of the search, DosMemAlloc() sets a local variable TRUE and terminates the loop, unless it determines that it encountered an end block after the test, in which case it returns an error.

Listing 5.2   The DosMemAlloc() function.

```
COUNT
DosMemAlloc (int size, COUNT mode, seg FAR *para, UWORD FAR *asize)
{
    REG mcb FAR *p;
    mcb FAR *q;
    COUNT i;
    BOOL found;

    /* Initialize                                              */
    p = (mcb FAR *)(MK_FP(first_mcb, 0));
```

For each iteration, `DosMemAlloc()` checks for a corrupted memory block. The test it performs simply looks at the `m_type` member to determine whether the block is valid. If it is, neither an ASCII "M" nor

Figure 5.2   Memory allocation criteria.

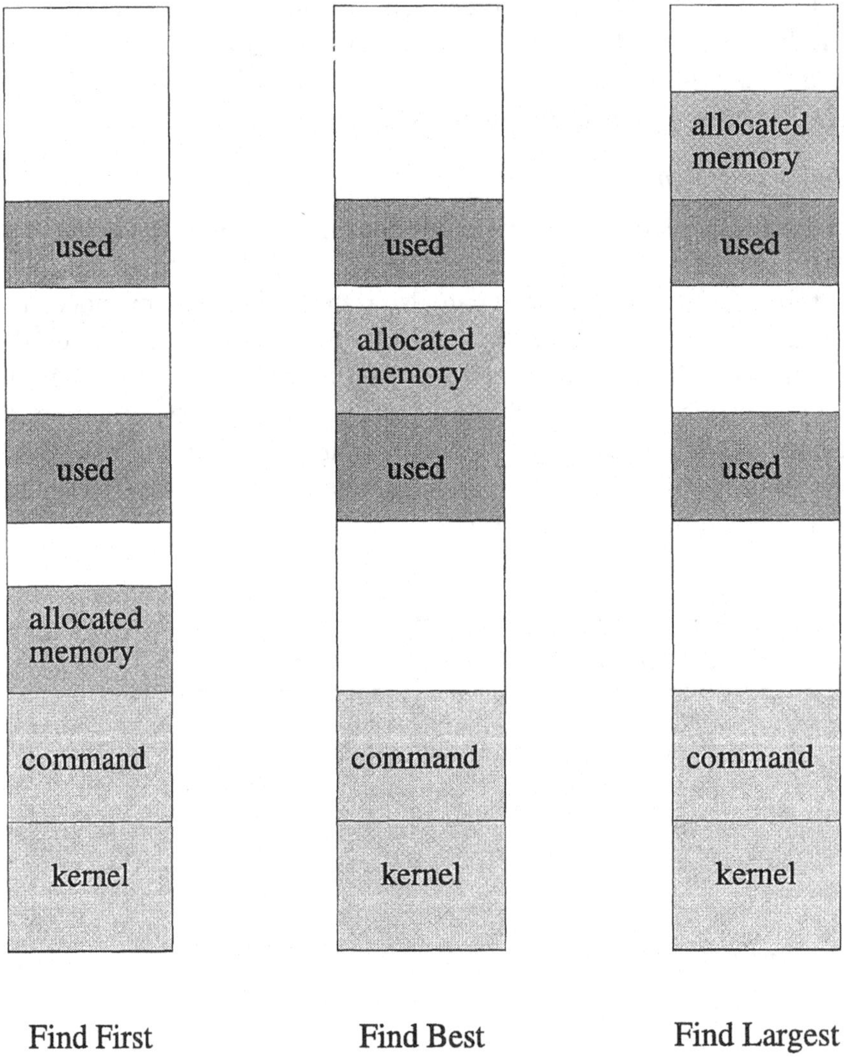

Listing 5.2   The DosMemAlloc() function — continued.

```
/* Search through memory blocks                          */
for(q = (mcb FAR *)0, i = 0, found = FALSE; !found; )
{
    /* check for corruption                              */
    if(p -> m_type != MCB_NORMAL && p -> m_type != MCB_LAST)
        return DE_MCBDESTRY;

    /* Test if free based on mode rules                  */
    switch(mode)
    {
    case LAST_FIT:
    default:
        /* Check for a last fit candidate                */
        if(p -> m_size >= size && p -> m_psp == FREE_PSP)
            /* keep the last know fit                    */
            q = p;
        /* not free - bump the pointer                   */
        if(p -> m_type != MCB_LAST)
            p = MK_FP(far2para((VOID FAR *)p) + p -> m_size +
                1, 0);
        /* was there no room (q == 0)?                   */
        else if(p -> m_type == MCB_LAST && q == (mcb FAR *)0)
            return DE_NOMEM;
        /* something was found - continue                */
        else
            found = TRUE;
        break;

    case FIRST_FIT:
        /* Check for a first fit candidate               */
        if(p -> m_size >= size && p -> m_psp == FREE_PSP)
        {
            q = p;
            found = TRUE;
            break;
        }
        /* not free - bump the pointer                   */
        if(p -> m_type != MCB_LAST)
            p = MK_FP(far2para((VOID FAR *)p) + p -> m_size +
                1, 0);
        /* nothing found till end - no room              */
        else
            return DE_NOMEM;
        break;
```

an ASCII "Z", DosMemAlloc() terminates and returns an error. If it is a valid memory block, DosMemAlloc() tests to see if it is free, and if so, proceeds to test it based on mode rules. The algorithm then diverges at this point when it tests the block. A switch statement splits the execution path based on the requested mode.

DosMemAlloc() tests for last fit by searching through each memory control block. If the size of the block is greater than or equal to the requested size, DosMemAlloc() notes the address of the block. If DosMemAlloc() examined the block marked with the end of the memory arena, it terminates and returns an error; otherwise, it proceeds to common completion code.

Listing 5.2   The DosMemAlloc() function — continued.

```
case BEST_FIT:
    /* Check for a best fit candidate                    */
    if(p -> m_size >= size && p -> m_psp == FREE_PSP)
    {
        if(i == 0 || p -> m_size < i)
        {
            i = p -> m_size;
            q = p;
        }
    }
    /* not free - bump the pointer                        */
    if(p -> m_type != MCB_LAST)
        p = MK_FP(far2para((VOID FAR *)p) + p -> m_size +
            1, 0);
    /* was there no room (q == 0)?                        */
    else if(p -> m_type == MCB_LAST && q == (mcb FAR *)0)
        return DE_NOMEM;
    /* something was found - continue                     */
    else
        found = TRUE;
    break;
```

Listing 5.2   The `DosMemAlloc()` **function — continued.**

```
        case LARGEST:
            /* Check for a first fit candidate                    */
            if((p -> m_psp == FREE_PSP) && (i == 0 || p -> m_size > i))
            {
                size = *asize = i = p -> m_size;
                q = p;
            }
            /* not free - bump the pointer                        */
            if(p -> m_type != MCB_LAST)
                p = MK_FP(far2para((VOID FAR *)p) + p -> m_size +
                    1, 0);
            /* was there no room (q == 0)?                        */
            else if(p -> m_type == MCB_LAST && q == (mcb FAR *)0)
                return DE_NOMEM;
            /* something was found - continue                     */
            else
                found = TRUE;
            break;
        }
    }
    p = q;
    /* Larger fit case                                            */
    if(p -> m_size > size)
    {
        if(mode != LAST_FIT)
        {
            q = MK_FP(far2para((VOID FAR *)p) + size + 1, 0);
            /* Always flow m_type up on alloc                     */
            q -> m_type = p -> m_type;
            p -> m_type = MCB_NORMAL;
            p -> m_psp = cu_psp;
            q -> m_psp = FREE_PSP;
            q -> m_size = p -> m_size - size - 1;
            p -> m_size = size;
            for(i = 0; i < 8; i++)
                p -> m_name[i] = q -> m_name[i] = '\0';
        }
    }
```

DosMemAlloc() tests for first fit in a way similar to the last fit case. It also searches through each memory control block. If the size of the block is greater than or equal to the requested size, the address of the block is noted. However, unlike last fit, it exits the loop and proceeds to common completion code. If DosMemAlloc() reaches the end of the memory arena, it also terminates and returns an error.

Listing 5.2   The DosMemAlloc() *function — continued.*

```
        else
        {
            q = MK_FP(far2para((VOID FAR *)p) +
                (p -> m_size - size), 0);
            /* Always flow m_type up on alloc               */
            q -> m_type = p -> m_type;
            p -> m_type = MCB_NORMAL;
            q -> m_psp = cu_psp;
            p -> m_psp = FREE_PSP;
            p -> m_size = p -> m_size - size - 1;
            q -> m_size = size;
            for(i = 0; i < 8; i++)
                p -> m_name[i] = q -> m_name[i] = '\0';
        }
        /* Found - return good                              */
        *para = far2para((VOID FAR *)(mode == LAST_FIT ?
                        (VOID FAR *)q : (VOID FAR *)p));
        return SUCCESS;
    }
    /* Exact fit case                                       */
    else if(p -> m_size == size)
    {
        p -> m_psp = cu_psp;
        for(i = 0; i < 8; i++)
            p -> m_name[i] = '\0';
        /* Found - return good                              */
        *para = far2para((VOID FAR *)(BYTE FAR *)p);
        return SUCCESS;
    }
    else
        return DE_MCBDESTRY;
}
```

The best fit test uses a local variable to maintain the size of the block closest to the requested size. In a manner similar to the last fit case, DosMemAlloc() compares the requested size to the block size. If the size of the block is greater than or equal to the requested size, DosMemAlloc() then proceeds to test the block size. If it is smaller than the last block that was greater than the requested size, DosMemAlloc() notes the address of the block. In this way, it determines the smallest block that will satisfy the request. If DosMemAlloc() reaches the end of the memory arena, it terminates and returns an error; otherwise, it proceeds to common completion code.

The test for the largest memory block is the simplest case. DosMemAlloc() simply determines which memory block is the largest and notes the address of this block. In this way, DosMemAlloc() determines the smallest block that will satisfy the request. If it does not find a block, it terminates and returns an error; otherwise, it proceeds to common completion code.

DosMemAlloc() completes the search by modifying the block that it found. If the block is larger than requested, DosMemAlloc() proceeds to carve out the size of block requested from the block found. If, however, the block found is identical in size to that requested, DosMemAlloc() simply reassigns the block. With all arena maintenance out of the way, DosMemAlloc() exits by returning the segment of paragraph following the MCB. This segment forms the address of the first byte of allocated memory that the user may utilize.

Memory Deallocation

DOS-C provides a complement to DosMemAlloc() with DosMemFree() (Listing 5.3). DosMemFree() takes the segment acquired from DosMemAlloc() and returns it to the free pool. In comparison, DosMemFree() is much simpler because there is only one way to return a memory block, whereas there are four ways to allocate one. Its job is a simple one: check that the block is valid and combine it with adjacent blocks if possible.

Listing 5.3   The `DosMemFree()` ***function.***

```
COUNT
DosMemFree (int para)
{
    REG mcb FAR *p, FAR *q;
    COUNT i;

    /* Initialize                                              */
    p = (mcb FAR *)(MK FP(para, 0));

    /* check for corruption                                    */
    if(p -> m_type != MCB_NORMAL && p -> m_type != MCB_LAST)
        return DE_INVLDMCB;

    /* Mark the mcb as free so that we can later               */
    /* merge with other surrounding free mcb's                 */
    p -> m_psp = FREE_PSP;
    for(i = 0; i < 8; i++)
        p -> m_name[i] = '\0';

    /* Now merge free blocks                                   */
    for(p = (mcb FAR *)(MK_FP(first_mcb, 0)); p -> m_type
            != MCB_LAST; p = q)
    {
        /* make q a pointer to the next block                  */
        q = MK_FP(far2para((VOID FAR *)p) + p -> m_size + 1, 0);
        /* and test for corruption                             */
        if(q -> m_type != MCB_NORMAL && q -> m_type != MCB_LAST)
            return DE_MCBDESTRY;
        if(p -> m_psp != FREE_PSP)
            continue;

        /* test if next is free - if so merge                  */
        if(q -> m_psp == FREE_PSP)
        {
            /* Always flow type down on free                   */
            p -> m_type = q -> m_type;
            p -> m_size += q -> m_size + 1;
            /* and make pointers the same since the next       */
            /* free is now this block                          */
            q = p;
        }
    }
    return SUCCESS;
}
```

DosMemFree() first checks for corruption in the same way that DosMemAlloc() does. DosMemFree() then marks the MCB as free so that you can later merge it with other surrounding free MCBs. Once marked, DosMemAlloc() proceeds to merge adjacent free blocks.

In order to free an MCB, DosMemFree() has to go through each one to determine if the previous block and the following block are available. It tries to merge all blocks that become available in a belt and suspenders fashion. The algorithm uses a pointer to the next block. DosMemFree() updates this pointer on each iteration based on the content of the last block and tests for corruption. Again, because of lack of memory protection hardware in low-end 80x86 processors, any task can corrupt an MCB.

If the MCB passes the corruption test, DosMemFree() tests the next MCB to see if it is free. If it is, it merges the two blocks by first setting the m_psp structure member to FREE_PSP. It then takes the type from what had been the adjacent free block and copies it into the current free block. The algorithm counts on the fact that only the last block is marked free; if you merge with the last block, then the current block becomes the last block. It then adjusts the size of the current block and exits.

The next function, DosMemLargest() (Listing 5.4), is never seen by the user. Its purpose in DOS-C is to assist the task manager in locating the largest available free block. Task manager needs this information because it follows the model of MS-DOS in its operation. Under certain conditions in the process of loading a program, the task manager must load the program in the largest available memory. DosMemLargest() is called to return this information.

Listing 5.4   The DosMemLargest() *function.*

```
seg
DosMemLargest (seg FAR *size)
{
    REG mcb FAR *p;
    mcb FAR *q;
    COUNT found;

    /* Initialize                                               */
    p = (mcb FAR *)(MK_FP(first_mcb, 0));
```

In a simple fashion, `DosMemLargest()` searches through each memory block to find the largest one. As you have seen before, first it checks for and exits if it finds arena corruption. It then tests for the largest block available by noting the segment of the current MCB. If it is larger than any you have encountered to this point, `DosMemLargest()` exits successfully; otherwise, it returns an error indicating that you are out of memory.

A user-accessible memory manager function is `DosMemChange()` (Listing 5.5), which changes the size of a given memory block, if possible. It can be used to either increase or decrease the size of the block. It can also be called without changing the size of the memory block, but that is a special case.

Listing 5.4   The `DosMemLargest()` function — continued.

```
/* Search through memory blocks                            */
for(q = (mcb FAR *)0, *size = 0, found = FALSE; !found; )
{
    /* check for corruption                               */
    if(p -> m_type != MCB_NORMAL && p -> m_type != MCB_LAST)
        return DE_MCBDESTRY;

    /* Test for largest fit/available                     */
    if((p -> m_psp == FREE_PSP) && (p -> m_size > *size))
    {
        *size = p -> m_size;
        q = p;
    }
    /* not free - bump the pointer                        */
    if(p -> m_type != MCB_LAST)
        p = MK_FP(far2para((VOID FAR *)p) + p -> m_size + 1, 0);
    /* was there no room (q == 0)?                        */
    else if(p -> m_type == MCB_LAST && q == (mcb FAR *)0)
        return DE_NOMEM;
    /* something was found - continue*/
    else
        found = TRUE;
}
if( q != 0)
    return SUCCESS;
else
    return DE_NOMEM;
}
```

Listing 5.5   The DosMemChange() function.

```
COUNT
DosMemChange (int para, int size)
{
    REG mcb FAR *p, FAR *q;
    REG COUNT i;

    /* Initialize                                                   */
    p = (mcb FAR *)(MK_FP(--para, 0));

    /* check for corruption                                         */
    if(p -> m_type != MCB_NORMAL && p -> m_type != MCB_LAST)
        return DE_MCBDESTRY;

    /* check for wrong allocation                                   */
    if(size > p -> m_size)
    {
        REG COUNT delta;

        /* make q a pointer to the next block                       */
        q = MK_FP(far2para((VOID FAR *)p) + p -> m_size + 1, 0);

        /* if next mcb is not free, error no memory                 */
        if(q -> m_psp != FREE_PSP)
            return DE_NOMEM;

        /* reduce the size of q and add difference to p             */
        /* but check that q is big enough first                     */
        delta = size - p -> m_size;
        if(q -> m_size < delta)
            return DE_NOMEM;
        q -> m_size -= delta;
        p -> m_size += delta;

        /* Now go back and adjust q, we'll make p new q             */
        p = MK_FP(far2para((VOID FAR *)q) + delta, 0);
        p -> m_type = q -> m_type;
        p -> m_psp  = q -> m_psp;
        p -> m_size = q -> m_size;
        for(i = 0; i < 8; i++)
            p -> m_name[i] = q -> m_name[i];

        /* and finished                                             */
        return SUCCESS;
    }
```

Like the other memory manager functions, DosMemChange() starts out by verifying that the memory block passed to it is valid. If it is not valid, it returns an error. If it is valid, it proceeds to check the three possible cases.

If the new size is larger than the current size, DosMemChange() checks the next MCB to see if the block is free. The next memory block must be free in order for DosMemChange() to allocate memory. Unlike other operating systems, DOS-C does not have access to memory management hardware to allocate physical memory to logical memory. It must take it from the next memory block. If the next block is not free, it returns an error.

Listing 5.5   The DosMemChange() function — continued.

```
/* else, shrink it down                                         */
else if(size < p -> m_size)
{
    /* make q a pointer to the new next block                   */
    q = MK_FP(far2para((VOID FAR *)p) + size + 1, 0);

    /* reduce the size of p and add difference to q             */
    q -> m_type = p -> m_type;
    q -> m_size = p -> m_size - size - 1;
    p -> m_size = size;

    /* Make certian the old psp is not last (if it was)         */
    p -> m_type = MCB_NORMAL;

    /* Mark the mcb as free so that we can later                */
    /* merge with other surrounding free mcb's                  */
    q -> m_psp = FREE_PSP;
    for(i = 0; i < 8; i++)
        q -> m_name[i] = '\0';

    /* now free it so that we have a complete block             */
    return DosMemFree(far2para((VOID FAR *)q));
}

/* otherwise, its a no-op                                       */
else
    return SUCCESS;
}
```

If the next block is free, DosMemChange() verifies that it is large enough to accommodate the memory change. If so, DosMemChange() proceeds to reduce the size of the next block and add the difference to the current block. Once DosMemChange() has reduced the size of the next block, it must move the MCB to the new location. With the new MCB in place, DosMemChange() successfully returns.

If DosMemChange() is to reduce the size of the memory block, it reduces the size of the current block and adds the difference to the next block. It creates a new MCB for the left-over memory and then exits through DosMemFree(), which will merge the freed block, if possible, completing the reallocation process.

The final possible scenario is when the new size is identical to the old size. In this case, DosMemChange() simply exits. Although there is no physical change, DosMemChange() does perform a validation of the memory block, This is a good way to check the validity of the block.

DosMemCheck() (Listing 5.6) is the final memory manager function I will examine. Like DosMemLargest(), it is not accessible to the user. It is a DOS-C function that supports the memory manager. Because of the critical nature of the MCB headers and the arena itself, DOS-C must

---

Listing 5.6   The DosMemCheck() function.

```
COUNT
DosMemCheck (void)
{
    REG mcb FAR *p;

    /* Initialize                                                */
    p = (mcb FAR *)(MK_FP(first_mcb, 0));

    /* Search through memory blocks                              */
    for( ; ; )
    {
        /* check for corruption                                  */
        if(p -> m_type != MCB_NORMAL && p -> m_type != MCB_LAST)
            return DE_MCBDESTRY;
```

check the validity of the arena at key times, such as when a program terminates. It is the responsibility of DosMemCheck() to guarantee that DOS-C does not operate with a corrupted memory arena.

The DosMemCheck() algorithm is simple. It begins at the beginning of the memory arena and examines the MCB header. It checks for corruption using the now familiar ASCII "M" or ASCII "Z" check. If the header is good, it computes the location of the next block and repeats the check. DosMemCheck() continues checking MCB headers until the final memory block has been checked.

Arena Support Functions

Aside from the arena management functions, the memory manager also supplies arena support functions. These functions are needed to support what may seem to be simple functions but are key to a portable design. For example, funtions are needed that extend the addition of scalar values to segmented pointers so that data references may extend past the 64Kb 80x86 boundaries. They are also important to assist in compiler independance for 80x86 processors. For example, integer arithmetic may wrap in one compiler or memory model and be adjusted in another.

Listing 5.6   The DosMemCheck() *function — continued.*

```
      /* not corrupted - if last we're OK!                    */
      if(p -> m_type == MCB_LAST)
          return SUCCESS;

      /* not corrupted - but not end, bump the pointer        */
      else if(p -> m_type != MCB_LAST)
          p = MK_FP(far2para((VOID FAR *)p) + p -> m_size + 1, 0);

      /* totally lost - bad exit                              */
      else
          return DE_MCBDESTRY;
  }
}
```

---

mcb_init() is a rather simple function (Listing 5.7) that initializes the memory arena. For real mode, this is rather simple. All you need to do is create a single memory block that is also the last block. From here on out, all the memory management functions maintain the MCB chains.

DOS-C also provides two support functions that perform a logical to physical translation — loosely speaking. Again, you only support real mode, and as a result, there really is not much in the way of logical to physical support. However, these functions are of considerable help in converting segmented architecture pointers to linear representations. You will see both these functions used throughout the task and memory managers.

The function far2para() (Listing 5.8) takes a far pointer and converts it to a segment address, which is different than simply taking the segment portion of the pointer. It computes a linear representation and then converts it back (Figure 5.3a). What you are left with is the largest segment possible for this address.

Why do you need this? The segmented architecture of the Intel family allows a pointer 0000:fff0h. This address is not convenient for a pointer because with most C compilers targeted for Intel processors, you can only increment this by 0fh before the pointer wraps around to 0000:0000h. Using far2para() for the same pointer, you build a new pointer of 0fff:0000h. Now you get a useful range of ffffh before

---

Listing 5.7   The mcb_init() function.

```
VOID
mcb_init (mcb FAR *mcbp, int size)
{
    COUNT i;

    mcbp -> m_type = MCB_LAST;
    mcbp -> m_psp = FREE_PSP;
    mcbp -> m_size = size;
    for(i = 0; i < 8; i++)
        mcbp -> m_name[i] = '\0';
    mem_access_mode = FIRST_FIT;
}
```

the pointer wraps around. Practically, you can guarantee a range of fff0h or 65,520 bytes before the pointer wraps around.

Function long2para() is much simpler but equally necessary (Listing 5.8). In many instances during program loading, a linear logical address is computed and long2para() converts this address to the

---

Listing 5.8   The far2para() and long2para() functions.

```
seg
far2para (VOID FAR *p)
{
    seg u1 = FP_SEG(p);
    offset u2 = FP_OFF(p);
    ULONG phy_addr;

    phy_addr = (((long)u1) << 4) + u2;
    return (phy_addr>>4);
}

seg
long2para (LONG size)
{
    return ((size + 0x0f)>>4);
}
```

Figure 5.3a   far to paragraph operation.

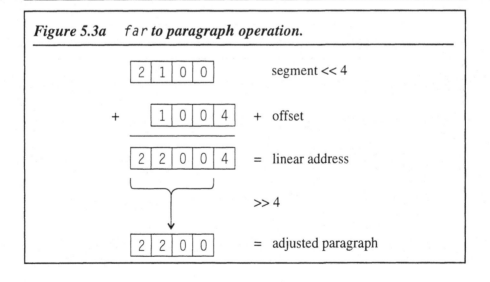

nearest segment (Figure 5.3b). Again, this function is a bridge between linear architecture and the segmented architecture of the Intel family.

Support functions with more complexity are add_far() and adjust_far() (Listing 5.9). Again, these functions are bridges between linear and segmented architectures. They also assist DOS-C in overcoming segmented pointer limitations.

As I discussed previously, many C compilers for Intel processors limit their pointer arithmetic to that which can be addressed within a segment. Unfortunately, DOS-C needs to address the entire 1Mb of real memory range. add_far() addresses this limitation. Both the file system manager and the memory manager need to add a long to a far pointer, with the result of a new far pointer. add_far() accomplishes

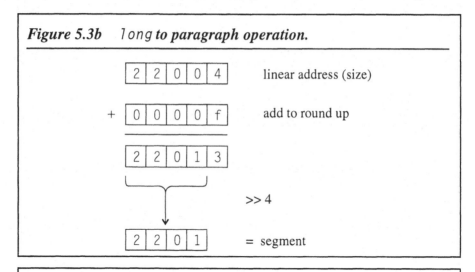

Figure 5.3b long *to paragraph operation.*

Listing 5.9 *The* add_far() *and* adjust_far() *functions.*

```
VOID FAR *
add_far (VOID FAR *fp, ULONG off)
{
    UWORD seg_val;
    UWORD off_val;

    /* Break far pointer into components                          */
    seg_val = FP_SEG(fp);
    off_val = FP_OFF(fp);
```

this by breaking a far pointer into components and adding the offset to the far pointer's offset part (Figure 5.4a). It then masks the offset so that it fits into a paragraph word and adds the top part into the segment. This top portion is effectively the carry into the segment. add_far() finally converts the individual components to a far pointer and returns it (Listing 5.9).

Listing 5.9 ***The*** add_far() ***and*** adjust_far() ***functions —***
continued.

```
    /* add the offset  to the fp's offset part            */
    off += off_val;

    /* make off_val equal to lower part of new value       */
    off_val = off & 0xffff;

    /* and add top part into seg                           */
    seg_val += ((off & 0x000f00001) / 0x10);

    /* and send back the new pointer                       */
    return (VOID FAR *)MK_FP(seg_val, off_val);
}

VOID FAR *
adjust_far (VOID FAR *fp)
{
    ULONG linear;
    UWORD seg_val;
    UWORD off_val;

    /* First, convert the segmented pointer to a linear address */
    linear = (((ULONG)FP_SEG(fp)) << 4) + FP_OFF(fp);

    /* Break it into segments.                             */
    seg_val = (UWORD)(linear >> 4);
    off_val = (UWORD)(linear & 0xf);

    /* and return an adddress adjusted to the nearest paragraph */
    /* boundary.                                           */
    return MK_FP(seg_val, off_val);
}
```

The function `adjust_far()` is also needed by the file system manager. It adjusts a `far` pointer so that you can get the maximum range out of C pointer arithmetic (Figure 5.4b), similar to `far2para()`. It does this by converting the segmented pointer to a linear address, breaking it into segments by masking, and returning an address adjusted to the nearest paragraph (Listing 5.9).

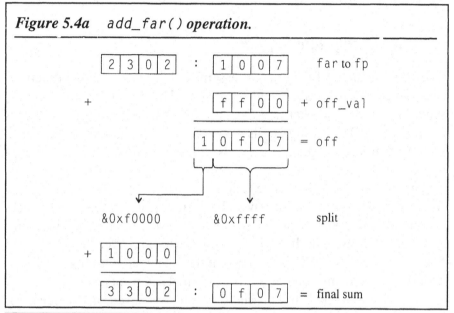

Figure 5.4a `add_far()` *operation.*

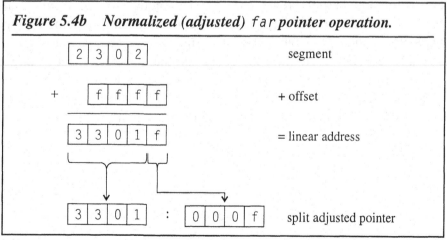

Figure 5.4b *Normalized (adjusted)* `far` *pointer operation.*

Task Manager

The DOS-C task manager is responsible only for starting tasks, switching between user space and kernel space, and terminating tasks. In contrast to other operating systems, it is a simple task manager; however, its simplicity will assist you in understanding machine support issues, such as initial register values and the switching of stack space.

Assembly Language Support

I will first examine fundamental assembly language support functions necessary for task management in DOS-C. These functions include the separation of user and kernel space, process start up, and process termination. The DOS-C design calls for a minimum of assembly language, so our assembly language functions are straightforward and not very complicated.

   The first assembly language function I will examine is exec_user( ) (Listing 5.10). This function saves the current state on the kernel stack. It then takes the pointer passed to it and recovers the new process context from it. It uses the Intel iret instruction to recover the cs:ip and flags from the stack. This is the simplest way of jumping to a new address that remains compatible with normal interrupts.

   You will notice that the processor context is saved on the stack for both user and kernel processes. This technique is used throughout DOS-C. There are other ways to save a context, such as placing all the registers within a table, but that requires much more code than using the stack. It does, however, place the constraint that the stack must be large enough for the user functions plus the context. Fortunately, MS-DOS addresses this in the programming manuals, and all programs written for MS-DOS will have a stack large enough to use this technique.

The same technique is used by the handle_break() function (Listing 5.10). DOS-C calls this function whenever a break has been detected. By MS-DOS standards, whenever a break is detected, an int 23h

---

Listing 5.10   The *exec_user( ) and* handle_break( ) *functions.*

```
;
;    Special call for switching processes
;
;    void interrupt far exec_user(irp)
;    iregs far *irp;
;
;    Borland C++ specific
;
;    +---------------+
;    |    irp hi  |    26
;    +---------------+
;    |    irp low |    24
;    +---------------+
;    |    flags  |    22
;    +---------------+
;    |    cs     |    20
;    +---------------+
;    |    ip     |    18
;    +---------------+
;    |    ax     |    16
;    +---------------+
;    |    bx     |    14
;    +---------------+
;    |    cx     |    12
;    +---------------+
;    |    dx     |    10
;    +---------------+
;    |    es     |    8
;    +---------------+
;    |    ds     |    6
;    +---------------+
;    |    si     |    4
;    +---------------+
;    |    di     |    2
;    +---------------+
;    |    bp     |    0
;    +---------------+
;
```

instruction is executed. This places the necessary cs:ip and flags for returning to the system on the user stack. When the user handler returns, it checks for an iret or retf return and handles each case appropriately.

Listing 5.10 *The* exec_user() *and* handle_break()
functions — continued.

```
                public      _exec_user
_exec_user      proc far

                push    ax
                push    bx
                push    cx
                push    dx
                push    es
                push    ds
                push    si
                push    di
                mov     ax,DGROUP
                mov     ds,ax
                push    bp
                mov     bp,sp
                cld
                cli
;
;
;
                mov     ax,WORD PTR [bp+24]     ; irp (user ss:sp)
                mov     dx,WORD PTR [bp+26]
                mov     sp,ax                  ; set-up user stack
                mov     ss,dx
                sti
;
        pop bp
                pop     di
                pop     si
                pop     ds
                pop     es
                pop     dx
                pop     cx
                pop     bx
                pop     ax
                iret

_exec_user              endp
```

Process Environment

The assembly language support functions you have just seen lay the foundation for the remainder of the task manager. These functions, along with memory manager support functions, are used extensively

---

Listing 5.10 **The** exec_user() **and** handle_break()
functions — continued.

```
;
;   Special call for switching processes during break handling
;
;   void interrupt far handle_break()
;
;
;   +---------------+
;   |    flags      |   24
;   +---------------+
;   |     cs        |   22
;   +---------------+
;   |     ip        |   20
;   +---------------+
;   |     ax        |   18
;   +---------------+
;   |     cx        |   16
;   +---------------+
;   |     dx        |   14
;   +---------------+
;   |     bx        |   12
;   +---------------+
;   |     sp        |   10
;   +---------------+
;   |     bp        |   8
;   +---------------+
;   |     si        |   6
;   +---------------+
;   |     di        |   4
;   +---------------+
;   |     ds        |   2
;   +---------------+
;   |     es        |   0    <--- bp & sp after mov bp,sp
;   +---------------+
;
```

within the loaders. However, before diving into the loaders, I want to look at a process environment.

A program, when loaded into memory, is a process. Each process within DOS-C has a data structure associated with it known as the PSP or Program Segment Prefix. It is important that you become familiar with this data structure because DOS-C uses it to tie together file system as well as task-related functions.

Listing 5.10   The exec_user() and handle_break()
functions — continued.

```
          public        _handle_break
_handle_break         proc far

          push  ax
          push  bx
          push  cx
          push  dx
          push  es
          push  ds
          push  si
          push  di
          mov   ax,DGROUP
          mov   ds,ax
          push  bp
          mov   bp,sp
          assume DS: DGROUP
          cld

          ; handler body - start out by restoring stack
          pushf
          cli

          ; save background stack
          mov   word ptr DGROUP:_api_ss,ss
          mov   word ptr DGROUP:_api_sp,sp

          ; restore foreground stack here
          mov   ss,word ptr DGROUP:_usr_ss
          mov   sp,word ptr DGROUP:_usr_sp
```

DOS-C also has two executable file types typically referred to as EXE and COM files. A COM file is a memory image of an executable program. DOS-C simply loads this type of program into memory. An EXE file, however, is a more complex file consisting of a header followed by the program and possibly a relocation table. It is important

---

Listing 5.10 **The** `exec_user()` **and** `handle_break()`
functions — continued.

```
                ; get all the user registers back
                pop     bp
                pop     di
                pop     si
                pop     ds
                pop     es
                pop     dx
                pop     cx
                pop     bx
                pop     ax

                ; do the int 23 handler and see if it returns
                int     23h

                ; we're back, must have been users handler
                push    ax
                push    bx
                push    cx
                push    dx
                push    es
                push    ds
                push    si
                push    di
                push    bp
                mov     bp,sp
                mov     ax,DGROUP
                mov     ds,ax
                assume DS: DGROUP

                ; test for far return or iret
                cmp     sp,_usr_sp
                jz      hbrk1              ; it was far ret
```

that you understand this data structure as well because it governs how the loader loads the program into memory.

As you have seen before, the psp data structure (Listing 5.11) is another 80x86 processor-dependent data structure. It is designed to fit within the first 16 paragraphs. This results in a few unused fill regions

---

Listing 5.10 **The** exec_user() **and** handle_break() **functions — continued.**

```
              ; restart int 21 from the top
hbrk1:        pop   bp
              pop   di
              pop   si
              pop   ds
              pop   es
              pop   dx
              pop   cx
              pop   bx
              pop   ax
              call  _int21_entry
              iret

hbrk2:        popf                      ; clear the flag from the stack
              jnc   hbrk1               ; user wants to restart
              pop   bp
              pop   di
              pop   si
              pop   ds
              pop   es
              pop   dx
              pop   cx
              pop   bx
              pop   ax
              mov   ax,4c00h            ; exit
                                        ; set break detected flag
              mov   byte ptr _break_flg,0ffh
              call  _int21_entry
              iret


_handle_break    edp
```

Listing 5.11   The *psp* **data structure.**

```
typedef struct
{
    UWORD  ps_exit;                      /* CP/M-like exit point        */
    UWORD  ps_size;                      /* memory size in paragraphs   */
    BYTE   ps_fill1;                     /* single char fill            */

    /* CP/M-like entry point                                            */
    BYTE   ps_farcall;                   /* far call opcode             */
    VOID   (FAR *ps_reentry)();          /* re-entry point              */
    VOID   (interrupt FAR *ps_isv22)(),  /* terminate address           */
           (interrupt FAR *ps_isv23)(),  /* break address               */
           (interrupt FAR *ps_isv24)();  /* critical error address      */
    UWORD  ps_parent;                    /* parent psp segment          */
    UBYTE  ps_files[20];                 /* file table - 0xff is unused */
    UWORD  ps_environ;                   /* environment paragraph       */
    BYTE FAR *ps_stack;                  /* user stack pointer - int 21 */
    WORD   ps_maxfiles;                  /* maximum open files          */
    UBYTE FAR *ps_filetab;               /* open file table pointer     */
    BYTE FAR *ps_prevpsp;                /* previous psp pointer        */
    BYTE FAR *ps_dta;                    /* process dta address         */
    BYTE   ps_fill2[16];
    BYTE   ps_unix[3];                   /* unix-style call - 0xcd 0x21 0xcb
*/
    BYTE   ps_fill3[9];
    union
    {
        struct
        {
            fcb
                _ps_fcb1;                /* first command line argument  */
        } _u1;
        struct
        {
            BYTE
                fill4[16];
            fcb
                _ps_fcb2;                /* second command line argument */
        } _u2;
        struct
        {
            BYTE  fill5[36];
            struct
            {
                BYTE _ps_cmd_count;
                BYTE _ps_cmd[127];       /* command tail                 */
            } _u4;
        } _u3;
    } _u;
} psp;
```

that are placed in the data structure to guarantee alignment and compatibility with MS-DOS. A few areas also serve multiple purposes. These areas result in unions becoming members of the structure. Although it may seem complex at first glance, studying each member will help in understanding the psp.

Historically, the psp has its roots in CP/M. The member ps_exit is a remnant of this CP/M heritage. It provides a CP/M-like exit point. A COM program can terminate by simply performing a jump to address 0. Another CP/M-like entry point is ps_farcall and ps_reentry(). These members are a call to the DOS-C system call dispatcher. A COM file may perform a system call with the call 5h mechanism that was used for CP/M. These two psp entries are now seldom used, but DOS-C maintains the structure members for compatibility with older programs.

Three members, ps_isv22(), ps_isv23(), and ps_isv24(), represent the initial process terminate address, break address, and critical error address vectors. Before DOS-C starts a process, the interrupt vectors for 22h, 23h, and 24h are copied into these psp structure members. Upon process termination, these vectors are copied back, and the 22h vector is executed. Quite often, these vectors are used as hooks into special termination processing.

DOS-C also uses the psp for process-related information. For example, there are system calls to set and get a quantity known as a DTA or Disk Transfer Area. Because this is a process-unique entity, it is stored within the psp in structure member ps_dta. Another per-process entity is ps_stack. This entry is the user stack pointer on entry into the int 21h handler. Other process-related members are ps_size, which holds the process memory size in paragraphs, and ps_parent, which contains the parent process psp segment. You should note that throughout DOS-C the psp segment is used as the process ID. It is this quantity that identifies the owner of a memory block and is stored in the MCB header.

I have previously noted that UNIX influenced MS-DOS in the area of file handles and redirection. psp data structure members ps_maxfiles, ps_files, and ps_filetab assist in the implementation. Member ps_maxfiles is a per-process variable that specifies the maximum number of open files permissible for this process. DOS-C uses a table for mapping user handles into system handles called ps_files. DOS-C also allows a user to change the number of files on a per-process basis. Member ps_filetab is a far pointer to the open file table.

Two other entries echo other UNIX influences on MS-DOS. In MS-DOS and DOS-C, each program has an environment that it inherits from its parent process. The implementation is a segment of memory aligned on a paragraph boundary. The segment for the environment is stored in ps_environ. Finally, a UNIX-style system call entry, ps_unix, allows a COM program to perform a call 50h to dispatch a system call.

The remainder of the psp is dedicated to two FCBs, _ps_fcb1 and _ps_fcb2, which are parsed prior to the start of the program. These are compatibility entries that were used for command line variables as a shortcut to processing the command tail. Of course, the command line used to invoke the process less the verb is _u4. These fcbs and the command tail are held within the union _u. This union is designed to specifically match the psp entries. Aliases for the union/struct combinations are _ps_fcb1 and _ps_fcb2 for the two system FCBs and _ps_cmd and _ps_cmd_count for the command tail.

Before becoming a process, a program may be stored on disk two ways. The first way is as a COM file, which is an exact binary image of the program as it initially appears in memory. The image must reserve the first 100h bytes for the psp and must use the tiny or small programming model. The second way is as an EXE file that is composed of a header, image, and an optional relocation table (Listing 5.12). The exe_header data structure is the mechanism used to tie the program together.

The first entry in `exe_header` is `exSignature`, which identifies an EXE file. This entry is either the ASCII characters "MZ" or "ZM". The next two members indicate the file size in 512-byte pages. The first of these two, `exExtraBytes`, indicates the number of bytes in the last partial page. The second of these two, `exPages`, contains the number of whole and partial pages in the file.

Member `exRelocItems` is a count of the number of relocation entries in the relocation table. Each entry is 4 bytes long and is stored in 80x86 segment:offset fashion. The final member that determines the EXE file size is `exHeaderSize`, which is the size of the header in 16-byte paragraphs. DOS-C uses this member to compute the size of the image that it loads into memory.

Before loading the program into memory, DOS-C attempts to allocate memory for the program. It does this by computing the image size and calculating minimum and maximum memory requirements using members `exMinAlloc` and `exMaxAlloc`. With memory allocated, the image is loaded, and the relocation table is used to fix segment references in the program. These segment references are pointed to by the

---

Listing 5.12   The `exe_header` data structure.

```
typedef struct
{
    UWORD exSignature;
    UWORD exExtraBytes;
    UWORD exPages;
    UWORD exRelocItems;
    UWORD exHeaderSize;
    UWORD exMinAlloc;
    UWORD exMaxAlloc;
    UWORD exInitSS;
    UWORD exInitSP;
    UWORD exCheckSum;
    UWORD exInitIP;
    UWORD exInitCS;
    UWORD exRelocTable;
    UWORD exOverlay;
} exe_header;
```

entries in the relocation table. DOS-C uses member exRelocTable to seek to the start of the relocation table within the file and then adjusts each segment reference for exRelocItems iterations.

With the program loaded into memory and far addresses adjusted for its new location, DOS-C uses exInitSS and exInitSP to set the initial stack pointer. It initializes each register and begins program execution at cs:ip specified by exInitIP and exInitCS.

Two final file maintenance entries remain in exe_header. The first is exOverlay, which is used for overlays and specifies the overlay number that a file represents. The other is exCheckSum, which is designed to contain a checksum that guarantees file integrity. Neither MS-DOS nor DOS-C uses this entry.

Now that I have covered the basic data structures, I can start looking at functions that fill these data structures. The DOS-C loaders use these functions to simplify the loading of executable programs.

The first function I will examine is ChildEnv() (Listing 5.13). This function creates a new environment for a process if one exists for the parent process. The first thing ChildEnv() does is compute the length of the environment by creating a far pointer and stepping through the strings until it finds the terminating null string. ChildEnv() attempts to protect the kernel against corrupted environments by testing the environment size previously computed to see if it is greater than the maximum permissible environment size. If it is, the environment may be corrupted or missing the terminating null string, and ChildEnv() aborts without creating a new environment area.

Listing 5.13   The ChildEnv() function.

```
COUNT
ChildEnv (exec_blk FAR *exp, UWORD *pChildEnvSeg)
{
    BYTE FAR *pSrc;
    BYTE FAR *pDest;
    UWORD nEnvSize;
    COUNT RetCode;
    UWORD MaxEnvSize;
```

Listing 5.13   The `ChildEnv()` *function — continued.*

```
    /* create a new environment for the process                        */
    if(exp -> exec.env_seg != 0)
    {
        for(nEnvSize = 0, pSrc = MK_FP(exp -> exec.env_seg,0);
            *pSrc != '\0'; )
        {
            while(*pSrc != '\0')
            {
                ++pSrc;
                ++nEnvSize;
            }
            /* account for terminating null                            */
            ++nEnvSize;
            ++pSrc;
        }
    }
    else
    {
/**     nEnvSize = 0; */
        pChildEnvSeg = 0;
        return SUCCESS;
    }

    /* Test env size and abort if greater than max                     */
    if(nEnvSize >= MAXENV)
        return DE_INVLDENV;

    /* allocate enough space for env + path                            */
    if((RetCode = DosMemAlloc(long2para(nEnvSize + 65), FIRST_FIT,
       (seg FAR *)pChildEnvSeg, (UWORD FAR *)MaxEnvSize)) < 0)
        return RetCode;
    else
        pDest = MK_FP(*pChildEnvSeg + 1, 0);

    /* fill the new env and inform the process of its                  */
    /* location throught the psp                                       */
    for(pSrc = MK_FP(exp -> exec.env_seg,0); *pSrc != '\0'; )
    {
        fstrncpy(pDest, pSrc, BUFFERSIZE);
        while(*pSrc != '\0')
            ++pSrc;
        while(*pDest != '\0')
            ++pDest;
        ++pSrc;
        ++pDest;
    }
    *pDest = '\0';
    return SUCCESS;
}
```

Now that you have the environment size and some assurance that it is intact, ChildEnv() calls DosMemAlloc() to allocate enough space for the new environment plus the path. If the memory is successfully allocated, ChildEnv() proceeds to fill the new environment and inform the new process of its location through the pointer pChildEnvSeg.

Just as the ChildEnv() function deals specifically with the creation of a process environment, DOS-C has the function new_psp() (Listing 5.14) that deals specifically with the creation of a process psp. Similar to its environment sibling, new_psp() is used by the loaders to simplify the loading of an executable program into memory. This design is not unusual. You have seen this type of design throughout DOS-C.

Listing 5.14   The new_psp( ) *function.*

```
VOID
new_psp (psp FAR *p, int psize)
{
    REG COUNT i;
    BYTE FAR *lpPspBuffer;

    /* Clear out new psp first                                  */
    for(lpPspBuffer = p, i = 0; i < sizeof(psp) ; ++i)
        *lpPspBuffer = 0;
    /* initialize all entries and exits                         */
    /* CP/M-like exit point                                     */
    p -> ps_exit = 0x20cb;
    /* CP/M-like entry point - jump to special entry            */
    p -> ps_farcall= 0xea;
#ifndef IPL
    p -> ps_reentry = int21_entry;
#endif
    /* unix style call - 0xcd 0x21 0xcb (int 21, retf)          */
    p -> ps_unix[0] = 0xcd;
    p -> ps_unix[1] = 0x21;
    p -> ps_unix[2] = 0xcb;

    /* Now for parent-child relationships                       */
    /* parent psp segment                                       */
    p -> ps_parent = DOS_PSP;
    /* previous psp pointer                                     */
    p -> ps_prevpsp = (BYTE FAR *)-1;
```

Listing 5.14   The `new_psp()` **function — continued.**

```
#ifndef IPL
    /* Environment and memory useage parameters              */
    /* memory size in paragraphs                             */
    p -> ps_size = psize;
    /* environment paragraph                                 */
    p -> ps_environ = 0;
    /* process dta                                           */
    p > ps_dta = (BYTE FAR *)(&p  > ps_cmd_count);
    /* terminate address                                     */
    p -> ps_isv22 = int22_handler;
    /* break address                                         */
    p -> ps_isv23 = int23_handler;
    /* critical error address                                */
    p -> ps_isv24 = int24_handler;
#endif

    /* File System parameters                                */
    /* user stack pointer - int 21                           */
    p -> ps_stack = (BYTE FAR *)-1;
    /* file table - 0xff is unused                           */
    for(i = 0; i < 20; i++)
        p -> ps_files[i] = 0xff;
    /* initialize stdin, stdout, etc                         */
    p -> ps_files[STDIN] = 0;     /* stdin                   */
    p -> ps_files[STDOUT] = 1;    /* stdout                  */
    p -> ps_files[STDERR] = 2;    /* stderr                  */
    p -> ps_files[STDAUX] = 3;    /* stdaux                  */
    p -> ps_files[STDPRN] = 4;    /* stdprn                  */
    /* maximum open files                                    */
    p -> ps_maxfiles = 20;
    /* open file table pointer                               */
    p -> ps_filetab = p -> ps_files;
    /* first command line argument                           */
    p -> ps_fcb1.fcb_drive = 1;
    /* second command line argument                          */
    p -> ps_fcb2.fcb_drive = 2;

    /* local command line                                    */
    p -> ps_cmd_count = 0;        /* command tail            */
    p -> ps_cmd[0] = 0;           /* command tail            */
}
```

The function new_psp( ) begins by clearing the memory region that will become the new psp. This action ensures a known state for members that are not currently used and also for variable-length fields such as the command tail.

The algorithm behind new_psp( ) is simple: it initializes all entries and exits. It does not fill in psp structure members that are process specific. In essence, it creates a skeletal psp that is necessary for both the COM loader and the EXE loader.

The first structure member it initializes is the CP/M-like exit point. It enters 2 bytes that are an int 20h instruction. This is the direct Terminate Program entry and guarantees termination whether the program performs a call or a jump to this location. It initializes the CP/M-like entry point next with a jump to special entry int21_entry. Then it initializes the UNIX-style call that is an int 21h instruction followed by a far return. This completes the psp system call linkages.

new_psp( ) next initializes the parent-child relationships. It fills in the parent psp segment with the DOS-C segment, DOS_PSP. This location will be overwritten later, but this initialization guarantees an owner. It then initializes the previous psp pointer to -1. This signifies to DOS-C that it is the start of the process chain. Again, it will be overwritten by DOS-C when applicable.

With process relationship structure members initialized, new_psp( ) proceeds to environment and memory usage structure members. It begins by initializing the process memory size in paragraphs with the parameter psize. Then, it initializes the environment with 0, indicating that no environment is assigned at this time. The program loaders will fill this in later. Next, new_psp( ) assigns the process dta to the command tail area, finishing initialization of the process memory structure member.

Next, new_psp( ) initializes the interrupt vector members. It first assigns the address of the int22_handler function to the terminate member. Then it initializes the break address member with the address of the int23_handler function. Finally, the critical error address structure member receives the address of the int24_handler function, and the user stack pointer member for int 21h system calls is initialized to -1. This last one is not a real vector but is grouped this way for simplicity.

new_psp() now proceeds to initialize the file system parameters. It first initializes the file table. Each entry is assigned 0xff, which indicates that the corresponding handle is unused or closed. It then initializes the first five handle entries for stdin, stdout, etc., with the corresponding table handles of the fixed system file. new_psp() then initializes the maximum open files structure member to the default of 20 and completes the file assignments by assigning the open file table pointer to the psp file table.

new_psp() completes its work by initializing the FCBs of the first and second command line arguments and the command tail, then exits. This completes the new_psp() function. Again, some of these entries will be overwritten, but this technique guarantees that the entire psp is initialized in one way or another and assists in the maintenance of DOS-C so that if any change accidentally omits a psp update, an error condition results that may be caught later.

Program Loaders

I now will take a look at the program loaders. This is the primary area of responsibility for the task loader of any DOS-compatible operating system. The basic methodology is simple. Look at the start of the file, if you see a magic number, assume that it is an EXE file; otherwise, it is a COM file. Based on this response, dispatch the correct loader.

The basic entry point for task loading is DosExec() (Listing 5.15). DosExec() begins by checking the validity of the file. First it looks for the file by simply trying to open the file with dos_open(). If DosExec() cannot find the file because dos_open() returns an error, it assumes that the file does not exist and returns an error. Next it attempts to read the header. If the file header cannot be read, DosExec() again assumes a nonexistent file and returns an error. If it was successful, DosExec() calls dos_close(rc) to close the file. Now for the acid test: if the header structure member exSignature is not MAGIC ("MZ"), then it assumes the file is a COM file and executes the COM loader; otherwise it assumes it's an EXE file and executes the EXE loader.

Listing 5.15   The `DosExec()` **function.**

```
COUNT
DosExec (COUNT mode, exec_blk FAR *ep, BYTE FAR *lp)
{
    COUNT rc, nNumRead;

    /* If file not found - free ram and return error          */
#ifdef IPL
    printf(".");
#endif
    if((rc = dos_open(lp, 0)) < 0)
        return DE_FILENOTFND;
    if((nNumRead = dos_read(rc, (VOID FAR *)&header,
                            sizeof(exe_header)))
      != sizeof(exe_header))
    {
#ifdef IPL
        printf("Internal IPL error - Read failure (read %d != %d)",
               nNumRead, sizeof(exe_header));
#endif
        return DE_INVLDDATA;
    }

    dos_close(rc);
#ifdef IPL
    printf(".");
#endif

    if(header.exSignature != MAGIC)
#ifndef IPL
        return DosComLoader(lp, ep, mode);
#else
    {
        char szFileName[32];

        fmemcpy((BYTE FAR *)szFileName, lp);
        printf("\nFile: %s (MAGIC = %04x)", szFileName,
               header.exSignature);
        fatal("IPL can't load .com files!");
    }
#endif
    else
        return DosExeLoader(lp, ep, mode);
}
```

The COM loader that `DosExec()` calls is `DosComLoader()` (Listing 5.16). `DosComLoader()` is a good function to examine first. By examining it, you can concentrate on the steps required to load a file into memory without the complexity of the EXE file relocation, placement in memory rules, etc.

Listing 5.16   The `DosComLoader()` function.

```
static COUNT
DosComLoader (BYTE FAR *namep, exec_blk FAR *exp, COUNT mode)
{
    COUNT rc, env_size, nread;
    UWORD mem;
    UWORD env, asize;
    BYTE FAR *sp, FAR *dp;
    psp FAR *p;
    mcb FAR *mp;
    iregs FAR *irp;
    LONG com_size;

    if(mode != OVERLAY)
    {
        if((rc = ChildEnv(exp, &env)) != SUCCESS)
            return rc;
    }
    else
        mem = exp -> load.load_seg;

    /* Allocate our memory and pass back any errors        */
    if((rc = DosMemAlloc(0, LARGEST, (seg FAR *)&mem,
                    (UWORD FAR *)&asize)) < 0)
    {
        if(env != 0)
            DosMemFree(env);
        return rc;
    }
    else
    {
        mp = MK_FP(mem, 0);
        ++mem;
    }
```

Listing 5.16   The DosComLoader() function — continued.

```
/* Now load the executable                                   */
/* If file not found - error                                 */
/* NOTE - this is fatal because we lost it in transit        */
/* from DosExec!                                             */
if((rc = dos_open(namep, 0)) < 0)
    fatal("(DosComLoader) com file lost in transit");
/* do it in 32K chunks                                       */
sp = MK_FP(mem, sizeof(psp));
for(com_size = 65536l, nread = 0; com_size > 0; )
{
    nread = dos_read(rc, sp, CHUNK);
    sp += nread;
    com_size -= nread;
    if(nread < CHUNK)
        break;
}
dos_close(rc);

if(mode == OVERLAY)
    return SUCCESS;

/* point to the PSP so we can build it                       */
p = MK_FP(mem, 0);
new_psp(p, mem + asize);

/* clone the file table                                      */
if(InDOS > 0)
{
    psp FAR *q;
    REG COUNT i;

    q = MK_FP(cu_psp, 0);
    for(i = 0; i < 20; i++)
    {
        REG COUNT ret;

        if(q -> ps_files[i] != 0xff
        && ((ret = CloneHandle(q -> ps_files[i])) >= 0))
            p -> ps_files[i] = ret;
        else
            p -> ps_files[i] = 0xff;
    }
}
```

As described earlier, a COM file is a simple memory image of the program. It is an exact image of the program at the instant prior to the start of execution. The rules are simple: reserve the first 256 bytes for a psp, and DOS-C guarantees to give you a stack, guarantees initial register values, and starts executing at 100h offset from the beginning of the file. This model is virtually identical to the CP/M model and made a good first choice for programs migrating from 8-bit to 16-bit environments.

Although the concept is simple, the algorithm is somewhat more complex. Some issues must be dealt with when starting a program. It needs to systematically allocate resources, make assignments to psp structure

---

Listing 5.16 **The** `DosComLoader()` **function — continued.**

```
p -> ps_environ = env == 0 ? 0 : env + 1;

/* complete the psp by adding the command line          */
p -> ps_cmd_count = exp -> exec.cmd_line -> ctCount;
fbcopy(exp -> exec.cmd_line -> ctBuffer, p -> ps_cmd,
        p -> ps_cmd_count > 127 ? 127 : p -> ps_cmd_count);

/* stick a new line on the end for safe measure         */
p -> ps_cmd[p -> ps_cmd_count] = '\r';

/* identify the mcb as this functions'                  */
/* use the file name less extension - left adjusted and */
/* space filled                                         */
mp -> m_psp = mem + 1;
for(asize = 0; asize < 8; asize++)
{
    if(namep[asize] != '.')
        mp -> m_name[asize] = namep[asize];
    else
        break;
}
for(; asize < 8; asize++)
        mp -> m_name[asize] = ' ';

/* build the user area on the stack                     */
irp = MK_FP(mem, (0xfffe - sizeof(iregs)));
```

members, and trap potential errors. Although these issues somewhat complicate the task of loading a COM file, the code is straightforward.

DosComLoader() begins by trying to allocate a user environment. It does this by calling ChildEnv() if it is not loading an overlay. DosComLoader() checks for errors and returns should ChildEnv() be unable to create the

---

Listing 5.16   The DosComLoader() function — continued.

```
/* start allocating REGS                                        */
irp -> ES = irp -> DS = mem;
irp -> CS = mem;
irp -> IP = 0x100;
irp -> AX = 0xffff; /* for now, until fcb code is in            */
irp -> BX =
irp -> CX =
irp -> DX =
irp -> SI =
irp -> DI =
irp -> BP = 0;
irp -> FLAGS = 0;

/* Transfer control to the executable                           */
p -> ps_parent = cu_psp;
p -> ps_prevpsp = (BYTE FAR *)MK_FP(cu_psp, 0);
p -> ps_stack = (BYTE FAR *)user_r;
switch(mode)
{
case LOADNGO:
    cu_psp = mem;
    dta = p -> ps_dta;
    exec_user(irp);
    /* We should never be here                                  */
    fatal("KERNEL RETURNED!!!");
    break;

case LOAD:
    exp -> exec.stack = (BYTE FAR *)irp;
    exp -> exec.start_addr = MK_FP(irp -> CS, irp -> IP);
    return SUCCESS;
}
}
```

environment. Given that the first memory allocation succeeds, it next tries to allocate memory for the process and exits if an error occurs. Note that by specification, the memory allocation is for the largest memory block available. Also, if DosComLoader() detects an error condition, it must deallocate the memory previously set aside for the environment. This pattern of deallocating previously set-aside resources continues through DosComLoader() because as more of the new process environment is built up, more system resources are used. An exit from an error condition must systematically return the system resources otherwise, DOS-C would slowly become useless after repeated errors.

At this point, DosComLoader() has allocated all the memory it needs to load the executable. It again starts out by attempting to open the file that contains the program image. If it cannot find the file, it performs error processing. Note that this time the error is fatal. The reason may not be clear until you examine how DOS-C arrived at this point. Previously, the file had to exist in order to test for the presence of a header. If an error occurs only a single function call later, a more serious error, such as the corruption of the user data or a wild function call, caused DOS-C to execute this code erroneously. The odds are very good that DOS-C cannot recover, so it must terminate.

Once DosComLoader() opens the file, it proceeds to load the image into memory in 32Kb chunks. It does this by repeatedly calling dos_read() and updating pointers so that the next iteration loads the next chunk in the proper location. There are only two ways to exit the loop: the first way is to try to read more than 64Kb, and the second way is to read less than 32Kb, indicating an end-of-file condition. Regardless of the terminating condition, DosComLoader() then closes the file by calling dos_close(). This is also the exit point for an overlay, since all other resources are part of the root program.

For program loading, however, DosComLoader() now must proceed to make this image into an executable process. It computes the location of the psp so that DosComLoader() can populate it. DosComLoader() calls new_psp() to initialize, then proceeds to clone the parent process' file table. DosComLoader() places the true environment segment value into the psp and then completes it by moving the command line into the command tail of the psp.

Next, DosComLoader() has to do some housekeeping. It needs to identify the mcb to signify that it belongs to this process. It does this in two ways. First, DosComLoader() uses the file name less extension after left adjusting it with trailing space. Second, it places the segment value of the psp address into the m_psp structure member. It is this structure member that DOS-C uses to identify free memory blocks.

Finally, DosComLoader() must pay attention to the needs to the processor itself. It first computes the location of the machine stack and proceeds to build the area on the stack. It builds this area by allocating register variables in a manner that complies with the exec_user() support function. This is how DosComLoader() guarantees the register contents. With the machine context initialized, DosComLoader() transfers control to the loaded executable image through exec_user(). One final note, if the function is invoked with a load and go condition, the exec_user() call should never return. If it does, you have another unrecoverable error and DosComLoader() invokes a fatal error.

You now have a better understanding, after examining DosComLoader(), of the basic steps in loading a program into memory and starting the process. I will now move on to DosExeLoader() (Listing 5.17) to examine the more complete EXE file-loading process.

Listing 5.17   The *DosExeLoader()* function.

```
static COUNT
DosExeLoader (BYTE FAR *namep, exec_blk FAR *exp, COUNT mode)
{
    COUNT rc, env_size, i, nBytesRead;
    UWORD mem, env, asize, start_seg;
    ULONG image_size;
    ULONG image_offset;
    BYTE FAR *sp, FAR * dp;
    psp FAR *p;
    mcb FAR *mp;
    iregs FAR *irp;
    UWORD reloc[2];
    seg FAR *spot;
    LONG exe_size;
#ifdef IPL
    BYTE szNameBuffer[64];
#endif
```

As with DosComLoader(), DosExeLoader() first attempts to clone the environment and create a memory arena for the new process. Unlike DosComLoader(), it next computes an image offset from the header because the image shares its size information within the file with the header and optional relocation and debugger information. With this information in hand, DosExeLoader() can now compute image size by adding the number of pages scaled to bytes to the size of the psp in bytes.

Listing 5.17   The *DosExeLoader( ) function — continued.*

```
#ifndef IPL
    /* Clone the environement and create a memory arena         */
    if(mode != OVERLAY)
    {
        if((rc = ChildEnv(exp, &env)) != SUCCESS)
            return rc;
    }
    else
        mem = exp -> load.load_seg;
#endif

    /* compute image offset from the header                     */
#ifdef IPL
    fscopy(namep, (BYTE FAR *)szNameBuffer);
    printf("\nEXE loader loading: %s", szNameBuffer);
#endif
    image_offset = (ULONG)header.exHeaderSize * 161;

    /* compute image size by removing the offset from the number */
    /* pages scaled to bytes      plus the remainder and the psp */
    /*  First scale the size                                    */
    image_size = (ULONG)(header.exPages) * 5121;
#if 0
    /* remove the offset                                        */
    image_size -= image_offset;
    /* add in the remainder bytes                               */
    if(header.exExtraBytes != 0)
        image_size -= (ULONG)(512 - header.exExtraBytes);
#endif
    /* and finally add in the psp size                          */
    if(mode != OVERLAY)
        image_size += (ULONG)long2para((LONG)sizeof(psp));
```

This is now the minimum size of the memory block necessary to load the file. It will be used later to determine the actual memory block size.

Next, DosExeLoader() attempts to find out how many paragraphs are available. It computes the minimum size of the EXE file by using the exMinAlloc header structure member. It then follows the rules for EXE file loading by attempting to allocate the maximum memory requested for the process — the exact amount available if less than the maximum requested but no less than the minimum requested. Once DosExeLoader() has determined the amount of memory to request from the memory manager, it allocates the memory and returns if any errors occur.

With a memory block available, DosExeLoader() next computes the far pointer to the space where the image will be loaded and proceeds to load the executable. In a manner similar to DosComLoader(), it attempts

---

Listing 5.17   The DosExeLoader() function — continued.

```
#ifndef IPL
    if(mode != OVERLAY)
    {
        /* Now find out how many paragraphs are available      */
        if((rc = DosMemLargest((seg FAR *)&asize)) != SUCCESS)
            return rc;
        else
        {
            exe_size = (LONG)long2para(image_size) +
                        header.exMinAlloc +
                        long2para((LONG)sizeof(psp));
            if(exe_size > asize)
                return DE_NOMEM;
            else if(((LONG)long2para(image_size) +
                    header.exMaxAlloc +
                    long2para((LONG)sizeof(psp))) < asize)
                exe_size = (LONG)long2para(image_size) +
                            header.exMaxAlloc +
                            long2para((LONG)sizeof(psp));
            else
                exe_size = asize;
        }
```

Listing 5.17   The *DosExeLoader( )* function — continued.

```
        /* Allocate our memory and pass back any errors      */
        /* We can still get an error on first fit if the above */
        /* returned size was a bet fit case                  */
        if((rc = DosMemAlloc((seg)exe_size, FIRST_FIT,
                        (seg FAR *)&mem,
                        (UWORD FAR *)&asize)) < 0)
        {
            if(rc == DE_NOMEM)
            {
                if((rc = DosMemAlloc(0, LARGEST, (seg FAR *)&mem,
                                (UWORD FAR *)&asize)) < 0)
                {
                    if(env != 0)
                        DosMemFree(env);
                    return rc;
                }
                /* This should never happen, but ...          */
                if(asize < exe_size)
                {
                    if(env != 0)
                        DosMemFree(env);
                    DosMemFree(mem);
                    return rc;
                }
            }
            else
            {
                if(env != 0)
                    DosMemFree(env);
                return rc;
            }
        }
        else
        /* with no error, we got exactly what we asked for    */
            asize = exe_size;
    }
#else
    mem = KERNSEG;
#endif
```

to open the file and exits with a fatal error if it cannot. The same logic that applied to DosComLoader() applies to DosExeLoader(); however, there is a major divergence at this point. Where DosComLoader() simply started loading from the beginning of the file, DosExeLoader() performs a seek operation from the offset computed earlier to the start of the image.

As in DosComLoader(), DosExeLoader() uses a loop to read in the image in 32Kb chunks. However, unlike DosComLoader(), its loop exit criteria are either reading all bytes in the computed EXE file size or a

---

Listing 5.17   The DosExeLoader() function — continued.

```
#ifndef IPL
    if(mode != OVERLAY)
        {
#endif
        /* memory found large enough - continue processing    */
        mp = MK_FP(mem, 0);
        ++mem;
#ifndef IPL
        }
    else
        mem = exp -> load.load_seg;
#endif

#ifdef IPL
    printf(".");
#endif
    /* create the start seg for later computations            */
    if(mode == OVERLAY)
        start_seg = mem;
    else
        start_seg = mem + long2para((LONG)sizeof(psp));

    /* Now load the executable                                */
    /* If file not found - error                              */
    /* NOTE - this is fatal because we lost it in transit     */
    /* from DosExec!                                          */
    if((rc = dos_open(namep, 0)) < 0)
        fatal("(DosExeLoader) exe file lost in transit");
```

Listing 5.17   The *DosExeLoader()* function — *continued.*

```
#ifdef IPL
    printf(".");
#endif
    /* offset to start of image                              */
    if (dos_lseek(rc, image_offset, 0) != image_offset)
    {
#ifndef IPL
        if(mode != OVERLAY)
        {
            DosMemFree(--mem);
            if(env != 0)
                DosMemFree(env);
        }
#endif
        return DE_INVLDDATA;
    }

#ifdef IPL
    printf(".");
#endif

    /* read in the image in 32K chunks                       */
    if(mode != OVERLAY)
        exe_size = image_size - long2para((LONG)sizeof(psp));
    else
        exe_size = image_size;
    sp = MK_FP(start_seg, 0x0);
    while(exe_size > 0)
    {
        nBytesRead = dos_read((COUNT)rc, (VOID FAR *)sp,
                    (COUNT)(exe_size < CHUNK ? exe_size : CHUNK));
        sp = add_far((VOID FAR *)sp, (ULONG)CHUNK);
        exe_size -= nBytesRead;
        if(nBytesRead == 0 || exe_size <= 0)
            break;
#ifdef IPL
        printf(".");
#endif
    }
```

read error. Also, DosExeLoader( ) calls add_far( ) to update the pointer to the next chunk. This call is necessary in order to avoid the pointer addition problems discussed earlier in the chapter. DosExeLoader( ), unlike DosComLoader( ), is not finished with the file when it completes the image load. It must next proceed to relocate the image for the new segment into which it was loaded. It does this by seeking to the offset given in the header structure member exRelocTable and looping the header structure member exRelocItems iterations. For each iteration, it calls dos_read( ) to read the relocation segment:offset file pointer,

---

Listing 5.17   The *DosExeLoader( )* function — *continued*.

```
#if 0
    /* Error if we did not read the entire image             */
    if(exe_size != 0)
        fatal("Broken exe loader (exe_size != 0)");
#endif

    /* relocate the image for new segment                    */
    dos_lseek(rc, (LONG)header.exRelocTable, 0);
    for (i=0; i < header.exRelocItems; i++)
    {
        if (dos_read(rc, (VOID FAR *)&reloc[0], sizeof(reloc))
            != sizeof(reloc))
        {
            return DE_INVLDDATA;
        }
        if(mode == OVERLAY)
        {
            spot = MK_FP(reloc[1] + mem, reloc[0]);
            *spot = *spot + exp -> load.reloc;
        }
        else
        {
        /*  spot = MK_FP(reloc[1] + mem + 0x10, reloc[0]);    */
            spot = MK_FP(reloc[1] + start_seg, reloc[0]);
            *spot = *spot + start_seg;
        }
    }
```

Listing 5.17   The `DosExeLoader()` *function — continued.*

```
#ifdef IPL
    printf(".");
#endif
    /* and finally close the file                          */
    dos_close(rc);

    /* exit here for overlay                               */
    if(mode == OVERLAY)
        return SUCCESS;

    /* point to the PSP so we can build it                 */
    p = MK_FP(mem, 0);
    new_psp(p, mem + asize);

#ifndef IPL
    /* clone the file table                                */
    if(InDOS > 0)
    {
        psp FAR *q;

        q = MK_FP(cu_psp, 0);
        for(i = 0; i < 20; i++)
        {
            REG COUNT ret;

            if(q -> ps_files[i] != 0xff
              && ((ret = CloneHandle(q -> ps_files[i])) >= 0))
                p -> ps_files[i] = ret;
            else
                p -> ps_files[i] = 0xff;
        }
    }

    p -> ps_environ = env == 0 ? 0 : env + 1;
#else
    p -> ps_environ = 0;
#endif
```

computes the corresponding memory location, and adds the segment value to the original segment value in memory.

When `DosExeLoader()` completes the file load and relocation process, it closes the file by calling `dos_close()`. As with `DosComLoader()`, this is also the exit point for an overlay.

In a manner similar to `DosComLoader()`, `DosExeLoader()` proceeds to make this image into an executable process by computing the location of and building the PSP.

Listing 5.17   The `DosExeLoader()` function — continued.

```
    /* complete the psp by adding the command line          */
    p -> ps_cmd_count = exp -> exec.cmd_line -> ctCount;
    fbcopy(exp -> exec.cmd_line -> ctBuffer, p -> ps_cmd,
        p -> ps_cmd_count > 127 ? 127 : p -> ps_cmd_count);

    /* stick a new line on the end for safe measure          */
    p -> ps_cmd[p -> ps_cmd_count] = '\r';

    /* identify the mcb as this functions'                   */
    /* use the file name less extension - left adjusted and  */
    /* space filled                                          */
    mp -> m_psp = mem + 1;
    for(i = 0; i < 8; i++)
    {
        if(namep[i] != '.')
            mp -> m_name[i] = namep[i];
        else
            break;
    }
    for(; i < 8; i++)
            mp -> m_name[i] = ' ';

    /* build the user area on the stack                      */
    irp = MK_FP(header.exInitSS + start_seg,
                ((header.exInitSP - sizeof(iregs)) & 0xffff));

#ifdef IPL
    printf(".\n");
#endif
```

Listing 5.17   The *DosExeLoader( )* function — continued.

```
    /* start allocating REGs                                        */
    /* Note: must match es & ds memory segment                      */
    irp -> ES = irp -> DS = mem;
    irp -> CS = header.exInitCS + start_seg;
    irp -> IP = header.exInitIP;
    irp -> AX = 0xffff;        /* for now, until fcb code is in     */
#ifdef IPL
    irp -> BX = BootDrive;
    irp -> CX = NumFloppies;
#else
    irp -> BX =
    irp -> CX =
#endif
    irp -> DX =
    irp -> SI =
    irp -> DI =
    irp -> BP = 0;
    irp -> FLAGS = 0;

    /* Transfer control to the executable                           */
    p -> ps_parent = cu_psp;
    p -> ps_prevpsp = (BYTE FAR *)MK_FP(cu_psp, 0);
    p -> ps_stack = (BYTE FAR *)user_r;
#ifdef IPL
    printf("Starting kernel ...\n");
#endif
    switch(mode)
    {
    case LOADNGO:
        cu_psp = mem;
        dta = p -> ps_dta;
        exec_user(irp);
        /* We should never be here                                  */
        fatal("KERNEL RETURNED!!!");
        break;

#ifndef IPL
    case LOAD:
        exp -> exec.stack = (BYTE FAR *)irp;
        exp -> exec.start_addr = MK_FP(irp -> CS, irp -> IP);
        return SUCCESS;
#endif
    }
}
```

DosExeLoader( ) also computes the location of the machine stack and proceeds to build the user area on the stack. Unlike DosComLoader( ), it uses information in the header to do this, whereas DosComLoader( ) simply determined the stack from the process environment it built. It also places the machine context on the stack and transfers control to the loaded executable image through exec_user( ) using the same logic as that used in DosComLoader( ). As you will see next, this common logic is necessary for an orderly return.

Returning From a Process

When a process completes, it must return to the calling process. However, it does not return directly but exits through DOS-C. For this reason, DOS-C provides a single function, return_user( ), to handle process termination. After you examine this function, you will better understand why the process context for starting a process requires a special stack format. You will also later see how this affects the system call interface.

The return_user( ) function (Listing 5.18) has two basic purposes. First, it must systematically tear down the process environment that DosComLoader( ) or DosExeLoader( ) built up. Second, it must recover the old process context and restore it. In doing this, DOS-C acts on behalf of the terminating process and control simply passes through DOS-C.

The first task return_user( ) undertakes is to close all files by looping through the file table and closing each file individually. Next, return_user( ) proceeds to return the parent process context. It starts this by restoring the parent process interrupt vectors that were previously saved in the psp. Next, return_user( ) begins to free all process memory if not a Terminate and Stay Resident (TSR) return. DOS-C sets a flag when a process returns through the TSR entry points. This flag is checked here to determine if memory resources and the psp itself should be discarded. If it is a TSR return, return_user( ) bypasses the return of memory; otherwise, it returns the environment first, then the memory block for the process itself.

Finally, return_user( ) reestablishes the parent process as the current process by restoring the process segment value as the current process and restoring the dta to the parent process' dta. With all this out of the way, return_user( ) finally reenters the parent at the point prior to the system call that started the process by calling exec_user( ).

Listing 5.18   The *return_user( )* function.

```
VOID
return_user (void)
{   psp FAR *p, FAR *q;
    REG COUNT i;

    /* Close all files                                          */
    for(i = 0; i < 20; i++)
        DosClose(i);

    /* restore parent                                           */
    p = MK_FP(cu_psp, 0);

    /* When process returns - restore the isv                   */
    setvec(0x22, p -> ps_isv22);
    setvec(0x23, p -> ps_isv23);
    setvec(0x24, p -> ps_isv24);

    /* And free all process memory if not a TSR return          */
    if(!tsr)
    {
        if(p -> ps_environ != (UWORD)0)
            DosMemFree(--p -> ps_environ);
        DosMemFree(--cu_psp);
    }
    cu_psp = p -> ps_parent;
    q = MK_FP(cu_psp,0);
    dta = q -> ps_dta;
    exec_user(p -> ps_stack);
}
```

## Chapter 6

DOS-C Kernel: API

If you are reading this book sequentially, you started looking at DOS-C from the bottom up. You studied the lowest levels of the I/O portions of the system by looking at the assembly language interface to the device drivers. You explored the buffer cache and the character canonical processing. You then discovered the relationships of the lower levels to the upper managers: the file system manager, the memory manager, and the task manager.

I am going to break that flow now and approach the system calls from the top down. Why the change you may ask? Up until this point, the internals were unfamiliar to you and there was no foundation to discuss the higher levels. Now you have a solid foundation in place, and you can begin building the structure on top by following the specifications (Figure 6.1).

I am now going to discuss the DOS APIs. Although there are a few, I am going to focus on two: int 20h and int 21h. These two represent the two methods used by programs to communicate with DOS and are the most interesting of the system calls.

Using this approach, the developer who is familiar with the API can easily relate the system call to what is going on under the hood. This approach also allows the developer who is not familiar with the API an opportunity

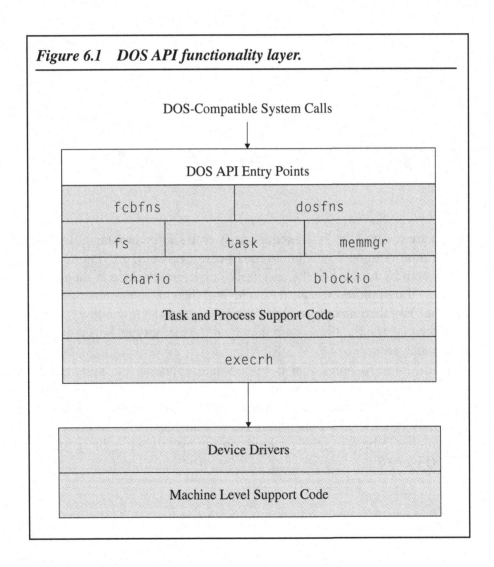

Figure 6.1   DOS API functionality layer.

to be guided through it and relate it to the internals. For developers in between those levels of familiarity, the following discussions will serve both as a refresher and guided tour of the interrelationships.

System Call Support

Warning: The code you are about to look at is highly 80x86 dependent. The reason is quite simple. Any code that deals with an operating system call is dependent on the underlying machine architecture. However, the methods used are portable to other processors if you generalize the technique.

Any processor must have some mechanism to make a call. For example, a Motorola 680x0 typically uses software exceptions. The Intel 80x86 uses two methods. The first method, the call gate, is valid only in protected mode. A particular memory location in the form of cs:ip is dedicated to the system call by placing an entry into a memory-resident table known as the translation lookaside buffer. The built-in memory management unit wakes up when a call is made to this address and causes the processor to dispatch the handler for it.

The second method, the software interrupt, is available in both real and protected modes. This method relies on a special instruction known as an int instruction. It functions in a way similar to a hardware interrupt or exception in that it causes a break in the normal flow of the code and forces the processor to begin a special piece of code to handle it. This instruction is usually followed by a number that indicates what vector address is to be used. To use it, you place the address of the handler into the table location that corresponds to the vector number used. You use it as a system call by making assignments to registers, or an area of memory, that are specially reserved for the system call, such as the stack. You then execute the int instruction, and when it returns, a register or memory location contains the results of the system call.

DOS-C and MS-DOS both use the software interrupt method of system calls of which there are two types. The first type is simply a call that takes no parameters. The Terminate call (int 20h) and the Share Processor call (int 28h) are representative of this. The second type is the more interesting of the two. This type of call takes both register and memory locations as variables and acts on them. In fact, the most interesting example is the int 21h call, where one register is used to multiplex between a number of system calls. It is this system call that gives DOS-compatible operating systems the bulk of their personalities.

Again, I warn the reader that the code you are about to study is very processor dependent. In fact, DOS-C uses the built-in interrupt function declarations that are common to 80x86 C compilers in order to simplify the design. However, you can use the same technique — by supplying an assembly language handler that acts as an interface to the C code — by saving the context onto the stack, setting the C call frame, performing the call, and then recovering the context before doing an interrupt return.

DOS-C has two simple system calls to handle program termination. One is the Terminate call where the program loses all its resources and returns to the parent program. The other is the Terminate and Stay Resident call where the program again returns to the parent program, but all memory resources remain intact. This type of program is the equivalent of a daemon process that is used in other operating systems. The differences between the two are minimal.

The function int20_handler() (Listing 6.1) is a very simple function terminate function. First, it sets the flag tsr to FALSE in order to differentiate between a normal exit and a TSR exit. This is the same flag you saw in the task manager in Chapter 5. It then calls DosMemCheck()

Listing 6.1   **The** int20_handler() *funtion.*

```
VOID INRPT far
int20_handler (int es, int ds, int di, int si, int bp, int sp,
               int bx, int dx, int cx, int ax, int ip, int cs,
               int flags)
{
    tsr = FALSE;
    DosMemCheck();
    return_user();
}
```

from the memory manager to guarantee that the memory arena is still intact since DOS-C cannot function reliably if the memory arena is corrupted. It finally calls return_user() from the task manager in order to terminate the task.

In the case of this system call, the only function an assembly language interface needs to perform is setting up the C call frame. As you can see from the function, none of the machine registers are used. However, this is not the case in the function int21_handler().

In DOS-C, the int 21h handler is divided into two functions (Figure 6.2). The first is int21_handler() (Listing 6.2). This is the function whose address is placed into the int 21h vector. It starts by incrementing the InDOS flag. This flag indicates to DOS-C that it

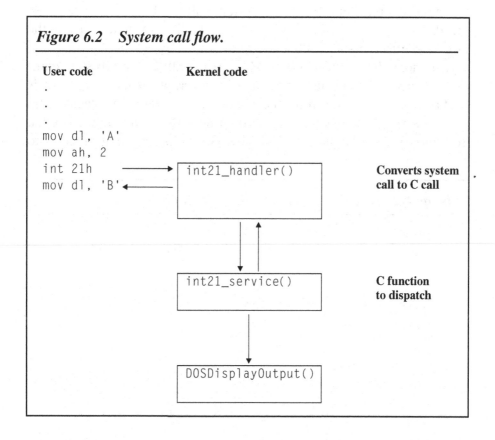

Figure 6.2   System call flow.

```
User code                 Kernel code

  .
  .
  .
mov dl, 'A'
mov ah, 2
int 21h         ────────▶  int21_handler()        Converts system
mov dl, 'B' ◀───────────                           call to C call


                           int21_service()        C function
                                                   to dispatch


                           DOSDisplayOutput()
```

received an interrupt from within the kernel. It causes some functions to execute differently in order to guarantee a certain degree of reentrant functionality. It next creates a pointer from the system call frame, user_r. This pointer is an important structure, iregs, that describes the interrupt stack frame and becomes the primary means of communication between the kernel and the user functions. All register references into and out of the kernel are contained on this stack. It is a far pointer because the context is contained within the executing program's stack space, again contributing to a reentrant design. With the machine context in place, int21_handler() switches from the user stack to the kernel stack by calling set_stack(). At this point, int21_handler() has created the kernel environment. It calls int21_service() to dispatch the function call and passes user_r, the pointer to the interrupt context, to allow the functions that were dispatched to access and modify the user's registers. When the system call has been completely serviced, int21_handler() calls restore_stack() to switch back to the user stack, decrements InDOS, and returns.

One note about this function. Most, if not all, of the 80x86 C compilers support the special function type interrupt. When a function is declared as a type interrupt, the compiler automatically places extra code to save all the machine registers. It does this in a particular order. To simplify the switching of tasks, DOS-C uses this register order to

---

Listing 6.2   The int21_handler() *function.*

```
VOID INRPT far
int21_handler (int es, int ds, int di, int si, int bp, int sp,
               int bx, int dx, int cx, int ax, int ip, int cs,
               int flags)
{
    ++InDOS;
    user_r = (iregs far *)&es;
    set_stack();
    int21_service(user_r);
    restore_stack();
    --InDOS;
}
```

save both user and kernel context, as well as the register context for starting a program (Figure 6.2). This allows a single structure definition for all process-related functions, including system calls. It also adds additional instructions to establish the data section for the function. With all this out of the way, it finally creates the stack frame for the C call and begins its work.

Listing 6.3   *A sample C interrupt function handler and assembly language equivalent.*

```
;
;   int21_handler (int es, int ds, int di, int si, int bp, int sp,
;                  int bx, int dx, int cx, int ax, int ip, int cs,
;                  int flags)
;
    assume          cs:_TEXT
_int21_handler      proc  far
    push  ax
    push  bx
    push  cx
    push  dx
    push  es
    push  ds
    push  si
    push  di
    push  bp
    mov   bp,DGROUP
    mov   ds,bp
    mov   bp,sp
;
;   {
;       ++InDOS;
;
    inc   word ptr DGROUP:_InDOS
;
;       user_r = (iregs far *)&es;
;
    lea   ax,word ptr [bp]
    mov   word ptr DGROUP:_user_r+2,ss
    mov   word ptr DGROUP:_user_r,ax
```

You will notice when examining the translated statements (Listing 6.3) that in the assignment of the user context `far` pointer, `user_r`, the compiler is aware that the registers are passed on the stack and uses `ss`

Listing 6.3   ***A sample C interrupt function handler and assembly language equivalent — continued.***

```
;
;           set_stack();
;
            call   near ptr _set_stack
;
;           int21_service(user_r);
;
    push   word ptr DGROUP:_user_r+2
    push   word ptr DGROUP:_user_r
    call   near ptr _int21_service
    pop    cx
    pop    cx
;
;           restore_stack();
;
    call   near ptr _restore_stack
;
;           --InDOS;
;
    dec    word ptr DGROUP:_InDOS
;
;       }
;
    pop    bp
    pop    di
    pop    si
    pop    ds
    pop    es
    pop    dx
    pop    cx
    pop    bx
    pop    ax
    iret
_int21_handler    endp
```

for the segment portion of the pointer. This assignment allows the stack to be switched in the very next statement with no adverse effects on the code.

The final aspect of `int21_handler()` to note is the exit. Upon terminating the function, the C compiler generates code to recover the context from the stack. This is the reason you needed to recover the user stack before returning. Also, the compiler generates an `iret` instead of the normal `ret` instruction to return from the interrupt. This makes the function a full interrupt handler in its own right, without ever having to resort to assembly language. These features may not be available for your C compiler or platform, but the code presented in Listing 6.2 should be used as a guide to your own specialized interrupt/system call handler.

Once safely within kernel space, `int21_service()` is the function that actually handles the system call (Listing 6.4). It is arranged as a `switch` statement based on the contents of the `ah` register. It accesses the `ah` register through the `iregs far` pointer `r` as `r -> AH`. This is important to note because all registers are accessed in this fashion and are part of the user context as explained earlier.

Listing 6.4   An abbreviated `int21_service()` function.

```
VOID int21_service(iregs far *r)
{
    COUNT rc, rc1;
    ULONG lrc;

    /* The dispatch handler                                        */
#ifdef DEBUG
    if(bDumpRegs)
    {
        fbcopy((VOID FAR *)r, (VOID FAR *)&error_regs,
               sizeof(iregs));
        printf("System call (21h): %02x\n", r -> AX);
        dump_regs = TRUE;
        dump();
    }
```

Listing 6.4   An abbreviated `int21_service()` **function —**
             continued.

```
#endif
    switch(r -> AH)
    {
    /* print unused and fall into terminate (debug only)        */
    default:
#ifdef DEBUG
        fbcopy((VOID FAR *)r, (VOID FAR *)&error_regs,
               sizeof(iregs));
        printf("Invalid system call (21h)\n");
        dump_regs = TRUE;
        dump();
#  ifdef KDB
        break;
#  endif
#else
        r -> AX = -rc;
        r -> FLAGS |= FLG_CARRY;
        break;
#endif

    /* Terminate Program                                         */
    case 0x00:
        tsr = FALSE;
        return_mode = break_flg ? 1 : 0;
        return_code = r -> AL;
        DosMemCheck();
        return_user();
        break;

    /* Read Keyboard with Echo                                   */
    case 0x01:
        r -> AL = DosCharInputEcho();
        break;

    /* Display Character                                         */
    case 0x02:
        DosDisplayOutput(r -> DL);
        break;

    .
    .
    .
```

The switch has a default handler up front. By specification, the carry flag is set on an invalid system call and the default case handles this condition. It also has a bit of debugging logic worked in to assist in determining DOS compatibility. It is switched in by defining DEBUG during compilation.

The rest of the switch is simple, although the cases are numerous. In each case, much of the manipulation of registers is done within the case

---

Listing 6.4   An abbreviated `int21_service()` **function —**
continued.

```
case 0x0f:
    if(FcbOpen(MK_FP(r -> DS, r -> DX)))
        r -> AL = 0;
    else
        r -> AL = 0xff;
    break;

case 0x10:
    if(FcbClose(MK_FP(r -> DS, r -> DX)))
        r -> AL = 0;
    else
        r -> AL = 0xff;
    break;
    .
    .
    .

case 0x14:
    {
        COUNT nErrorCode;

        if(FcbRead(MK_FP(r -> DS, r -> DX), &nErrorCode))
            r -> AL = 0;
        else
            r -> AL = nErrorCode;
        break;
    }
    .
    .
    .
```

itself. The case sets up either to complete the entire call itself, such as in the CP/M compatibility functions 0x18, 0x1d, 0x1e, and 0x20, or to make a function call to a handler for that system call. Sometimes it requires setting up local variables, at which time the case is a compound statement. In almost all cases, the case exits through a break but not until it performs some error processing. Unfortunately, the error processing must be done within the case because MS-DOS has different error returns for groups of system calls.

Listing 6.4   An abbreviated `int21_service()` **function — continued.**

```
/* CP/M compatibility functions                              */

case 0x18:
case 0x1d:
case 0x1e:
case 0x20:
    r -> AL = 0;
    break;

    .
    .
    .

/* Parse File Name                                           */
case 0x29:
    {
        BYTE FAR *lpFileName;

        lpFileName = MK_FP(r -> DS, r -> SI);
        r -> AL = FcbParseFname(r -> AL,
         &lpFileName,
         MK_FP(r -> ES, r -> DI));
        r -> DS = FP_SEG(lpFileName);
        r -> SI = FP_OFF(lpFileName);
    }
    break;

    .
    .
    .
```

Examine a few of the system calls. Perhaps the simplest function call that dispatches into the kernel is Read Keyboard with Echo, function 0x01. In this function call, the al register returns a character entered at the keyboard after echoing it to the display. DOS-C provides the function DosCharInputEcho() to perform this function. The service code for the system call is a simple call to DosCharInputEcho() and an assignment of the return to r -> AL. Because no error is returned by this system call, it just terminates with a break to end system call processing.

Another example with more functionality is the Display Character call, function 0x02. For this function, the character to be displayed is passed in the dl register. To provide this functionality, DOS-C provides the function call DosDisplayOutput(), which takes a single parameter: the character to be displayed. The call is invoked with the argument r -> DL in order to retrieve the character from the user's dl register.

Listing 6.4 An abbreviated `int21_service()` *function —* **continued.**

```
/* Dos Open                                                */

case 0x3d:
    if((rc = DosOpen(MK_FP(r -> DS, r -> DX), r -> AL)) < 0)
    {
        r -> AX = -rc;
        r -> FLAGS |= FLG_CARRY;
    }
    else
    {
        r -> AX = rc;
        r -> FLAGS &= ~FLG_CARRY;
    }
    break;
```

Listing 6.4   An abbreviated `int21_service()` *function —*
**   continued.**

```
    /* Dos Close                                           */
    case 0x3e:
        if((rc = DosClose(r -> BX)) < 0)
        {
            r -> AX = -rc;
            r -> FLAGS |= FLG_CARRY;
        }
        else
            r -> FLAGS &= ~FLG_CARRY;
        break;

    /* Dos Read                                            */

    case 0x3f:
        rc = DosRead(r -> BX, r -> CX, MK_FP(r -> DS, r -> DX),
                     (COUNT FAR *)&rc1);
        if(rc1 != SUCCESS)
        {
            r -> FLAGS |= FLG_CARRY;
            r -> AX = -rc1;
        }
        else
        {
            r -> FLAGS &= ~FLG_CARRY;
            r -> AX = rc;
        }
        break;
.
.
.
#ifdef DEBUG
    if(bDumpRegs)
    {
        printf("Exiting system call (21h)\n");
        fbcopy((VOID FAR *)r, (VOID FAR *)&error_regs,
               sizeof(iregs));
        dump_regs = TRUE;
        dump();
        printf("---\n");
    }
#endif
}
```

A somewhat more complicated system call is Parse File Name, function 0x29. In this function, the filename to be parsed is at the address given by ds:si in the user registers, and the parsing is controlled by a flag in the al register. The function returns the next byte in the user's ds and si registers, after the parse terminates, and a result flag in the al register. In this example, you have local storage in the form of lpFileName and data transfers between internal functions and user registers. This type of function is much more common throughout DOS-C.

Our final example requires looking at two function calls, Dos Open, function 0x3d, and FCB Open, function 0x0f. These functions illustrate the logic behind two of the error returns in DOS-C. Both functions retrieve their arguments from user registers in a manner already seen. However, both functions have error returns. For FcbOpen(), a false return indicates an error. However, the system call returns an error condition in al. In order to accommodate, the case examines the return from the function call and assigns the correct value using an if...else construct. However, the function DosOpen() returns not only an error condition by returning a negative number, but returns a handle, which is a positive integer for successful completion. The specifications for this function call require that ax contains either the handle or an error number. The carry flag is checked to detect an error. This technique is much more common for the later system calls. Again, the case makes a call to DosOpen() and uses an if...else construct to build the correct return. However, there are several differences between FcbOpen() and DosOpen(). The DosOpen() test within the if statement is for negative numbers. Also, the returned value is always in the ax register and accessed through r -> AX, although the sign of the error must be changed. Additionally, the carry flag is definitively set or cleared.

Although these are the highlights of int21_service(), the system calls examined provide insight into the steps required to convert an int 21h call into a C function and vice versa (Figure 6.2). Unfortunately, each system call must be handled in a unique manner because of the way MS-DOS handles each call. But the overhead is minimal and the resulting compatibility far outweighs the headaches. Other system calls are a variation of the two examined here and you are encouraged to study them and expand on them for your own applications.

DOS-C Personality Layer

If you have been following the system calls, you will notice that they call functions that you have not yet encountered. The reason is that there are functionality layers between the core DOS-C functions and the system calls (Figure 6.3). These functionality layers fall into three categories: simple I/O, FCB calls, and DOS calls. I will examine each of these three and study the relationships between the system calls and the core DOS-C functions.

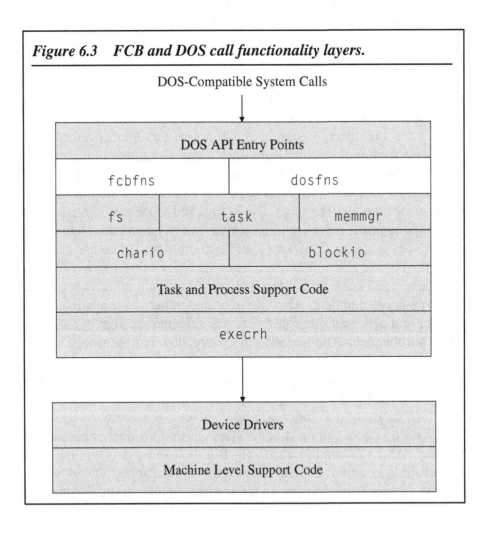

Figure 6.3   FCB and DOS call functionality layers.

Simple I/O Support

Start by looking at the simple I/O functions. A number of them support the console, an auxiliary device, and a printer. These are direct calls down to either the device drivers or a DOS support function. They may not need translation and are sometimes simply links from the int 21h system call to the device driver.

The function DosCharInputEcho() (Listing 6.5) is an example of simple I/O functions. It builds a request in the local variable rq, calls execrh(), and handles errors through char_error() should any occur. It then exits but not before calling DosCharOutput() to echo the character.

The function DosDisplayOutput() (Listing 6.6) is another example, although somewhat more complex. This function does some output formatting. It checks for a return, delete, backspace, and horizontal tab. In each case except tab, it updates internal screen parameters then calls DosCharOutput() to output the character to the screen. Tab repeatedly calls DosCharOutput() to output a space until the screen position variable scr_pos is a multiple of eight. This algorithm is how DOS-C performs tab expansion.

Listing 6.5   The DosCharInputEcho() function.

```
int DosCharInputEcho(VOID)
{
    BYTE cb;
    request rq;

    rq.r_length = sizeof(request);
    rq.r_command = C_INPUT;
    rq.r_count = 1;
    rq.r_trans = (VOID FAR *)&cb;
    rq.r_status = 0;
    execrh((request FAR *)&rq, syscon);
    if(rq.r_status & S_ERROR)
        return char_error(&rq, con_name);
    DosCharOutput(cb);
    return cb;
}
```

There are other simple I/O functions, but by now you should understand how these functions work. I urge you to look at the file inthndlr.c to further study these functions.

Listing 6.6   The `DosDisplayOutput()` **function.**

```
VOID DosDisplayOutput(COUNT c)
{
    /* Non-portable construct                                 */
    if(c < ' ' || c == 0x7f)
    {
        switch(c)
        {
        case '\r':
            scr_pos = 0;
            break;

        case 0x7f:
            ++scr_pos;
            break;

        case '\b':
            if(scr_pos > 0)
                --scr_pos;
            break;

        case '\t':
            do
                DosCharOutput(' ');
            while(scr_pos & 7);
            return;

        default:
            break;
        }
        DosCharOutput(c);
    }
    else
    {
        DosCharOutput(c);
    }
}
```

DOS Function Support

The next logical area to examine is the DOS Function Support, which includes functions that give DOS system calls, such as Make Directory, Open, etc., their functionality. These system calls typically work with a structure known as a system file table (SFT). Each entry is defined as the data structure sft in DOS-C.

The sft data structure (Listing 6.7) is key to both DOS and File Control Block (FCB) support functions. Although DOS-C does not use

---

Listing 6.7   The sft data structure.

```
typedef struct
{
    WORD  sft_count;         /* reference count                */
    WORD  sft_mode;          /* open mode - see below          */
    BYTE  sft_attrib;        /* file attribute - dir style     */
    WORD  sft_flags;         /* flags - see below              */
    union
    {
        struct      dpb FAR *
            _sft_dcb;        /* The device control block       */
        struct dhdr FAR *
            _sft_dev;        /* device driver for char dev     */
    } sft_dcb_or_dev;
    WORD  sft_stclust;       /* Starting cluster               */
    time  sft_time;          /* File time                      */
    date  sft_date;          /* File date                      */
    LONG  sft_size;          /* File size                      */
    LONG  sft_posit;         /* Current file position          */
    WORD  sft_relclust;      /* File relative cluster          */
    WORD  sft_cuclust;       /* File current cluster           */
    WORD  sft_dirdlust;      /* Sector containing cluster      */
    BYTE  sft_diridx;        /* directory index                */
    BYTE  sft_name[11];      /* dir style file name            */
    BYTE  FAR *
            sft_bshare;      /* backward link of file sharing sft */
    WORD  sft_mach;          /* machine number - network apps  */
    WORD  sft_psp;           /* owner psp                      */
    WORD  sft_shroff;        /* Sharing offset                 */
    WORD  sft_status;        /* this sft status                */
} sft;
```

sft for file operations, it maintains it for compatibility with MS-DOS. Although sft is well known and used by many DOS applications, it is one of the "undocumented" data structures. It is useful to explain the SFT to better understand the DOS and FCB support functions.

There is one entry for each open file in the system. This is the same as MS-DOS. Where DOS-C differs from MS-DOS is that it links an f_node to each SFT entry. This allows us to use the file system manager in place of DOS-specific file system code.

Review the sft so you better understand how to use it within the interface level. Each structure starts with the member sft_count, a reference count. This entry represents the number of file handles that refer to the file. In DOS-C, you maintain this entry and declare the SFT entry free when sft_count contains 0. This is the only real structure member that is used for SFT maintenance. The rest of the structure is devoted to the file.

The first of the file-related members, sft_mode, contains the mode the file was opened in (e.g., read, write, etc.). It is identical to the mode word passed as an argument to the DosOpen() system call (int 21h/function 3dh). The attribute of the file, as stored in the directory, is stored in the structure member sft_attrib. In DOS-C this entry is updated for application purposes only because the real file attribute is derived from the f_node. Also associated with file functionality is sft_flags, which holds a set of flags that determine the attributes of the open file. If the file is a character device, this member determines whether it is a console or null device, or whether the file should return an EOF condition. If it is a block device, it contains information such as the block device number and network file or drive information.

Following sft_flags is sft_dcb_or_dev. This union holds the far pointer _sft_dcb, which points to the device control block, and the far pointer _sft_dev, which points to the device driver for a character device. DOS-C uses this member whenever the file is a character device. For block devices, it maintains the field but uses the f_node information instead. Again, although DOS-C does not actively use an entry, it is important that the entry is maintained for compatibility.

The same is true for a number of file-related entries in the sft. These members, which I will briefly discuss next, are not used by DOS-C but are maintained for compatibility. They are sft_stclust, sft_time, sft_date, sft_size, sft_posit, sft_relclust, sft_cuclust, sft_dirdlust, sft_diridx, and sft_name. These members all have a corresponding entry within the f_node.

The remaining members are used for remote and shared applications. Both members sft_shroff and sft_bshare are used by share.exe for file-sharing purposes. sft_bshare contains the backward link of the file-sharing SFT, and sft_shroff holds the sharing offset. The remaining entries, with the exception of sft_status and sft_psp, deal with network files. The member sft_mach is used to hold the machine number for network applications.

In order to identify the owner of the file, sft_psp contains the Program Segment Prefix (PSP) of the owner. This field is actively used by DOS-C for the same purposes. Finally, sft_status contains the status of this SFT. In the case of DOS-C, it is actually the index of the corresponding f_node in the f_node table and is the link between the sft and the f_node.

With the knowledge of the underlying sft data structure, now look at the DOS personality functions. The first function you will look at is DosOpen(). It is representative of the personality functions that allocate an SFT. The method you will use is the same as you have used before: follow a typical sequence of open, read, and close and look at the details along the way.

DosOpen() (Listing 6.8) starts by checking for valid arguments. As you have seen before, this is a necessary step to prevent the propagation of an error that may crash the system. For this function, the only test is a check to see if the mode argument is within range.

Next, DosOpen() proceeds to open the file. First it must get a free handle. It checks the return to make certain that the handle was actually allocated. If not, it returns an error. Once it gets a handle, it has to get an SFT entry, again checking for an error.

Once resources have been allocated, DosOpen() must check for a device. The way DOS-C handles the file is dependent on whether or not the name specifies a disk file or a device. If it is a device, the sft is

---

Listing 6.8   The DosOpen() function.

```
COUNT
DosOpen (BYTE FAR *fname, COUNT mode)
{
    psp FAR *p = MK_FP(cu_psp,0);
    WORD hndl, sft_idx;
    sft FAR *sftp;
    struct dhdr FAR *dhp;
    BYTE FAR *froot;
    BYTE buff[FNAME_SIZE+FEXT_SIZE];
    WORD i;

    /* test if mode is in range                              */
    if((mode & SFT_OMASK) != 0)
        return DE_INVLDACC;

    /* get a free handle                                     */
    if((hndl = get_free_hndl()) == 0xff)
        return DE_TOOMANY;

    /* now get a free system file table entry                */
    if((sftp = get_free_sft((WORD FAR *)&sft_idx)) == (sft FAR *)-1)
        return DE_TOOMANY;

    /* check for a device                                    */
    froot = get_root(fname);
    for(i = 0; i < FNAME_SIZE; i++)
    {
        if(*froot != '\0' && *froot != '.')
            buff[i] = *froot++;
        else
            break;
    }
```

Listing 6.8   The DosOpen() function — continued.

```c
    for( ; i < FNAME_SIZE; i++)
        buff[i] = ' ';

    /* if we have an extension, can't be a device         */
    if(*froot != '.');
    {
        for(dhp = (struct dhdr FAR *)&nul_dev; dhp !=
            (struct dhdr FAR *)-1; dhp = dhp -> dh_next)
        {
            if(fnmatch((BYTE FAR *)buff, (BYTE FAR *)dhp ->
                dh_name, FNAME_SIZE, FALSE))
            {
                sftp -> sft_count += 1;
                sftp -> sft_mode = mode;
                sftp -> sft_attrib = 0;
                sftp -> sft_flags =
                    (dhp -> dh_attr & ~SFT_MASK) | SFT_FDEVICE |
                      SFT_FEOF;
                sftp -> sft_psp = cu_psp;
                fbcopy((BYTE FAR *)buff, sftp -> sft_name,
                        FNAME_SIZE+FEXT_SIZE);
                sftp -> sft_dev = dhp;
                p -> ps_filetab[hndl] = sft_idx;
                return hndl;
            }
        }
    }
    sftp -> sft_status = dos_open(fname, mode);
    if(sftp -> sft_status >= 0)
    {
        p -> ps_filetab[hndl] = sft_idx;
        sftp -> sft_count += 1;
        sftp -> sft_mode = mode;
        sftp -> sft_attrib = 0;
        sftp -> sft_flags = 0;
        sftp -> sft_psp = cu_psp;
        fbcopy((BYTE FAR *)buff, sftp -> sft_name,
                FNAME_SIZE+FEXT_SIZE);
        return hndl;
    }
    else
        return sftp -> sft_status;
}
```

initialized and the member `sft_flags` is set to indicate this. Also, the member `sft_dev` is initialized to point to the character device driver. This is the switch between character and file types. Device independence is also achieved by this, since the entry and all following functions deal with the driver in an identical fashion.

If the name did not specify a device, `DosOpen()` proceeds to call `dos_open()` to open the disk file. The return from this call is stored in `sft_status` and tested for an error. If no error occurred, this entry is now the link between the `sft` and the `f_node`. For a successful disk file open, the `sft` is initialized, and the function returns the handle for this open.

Once a file is opened you can proceed to read from it. The DOS personality function that supports the system call is `DosRead()` (Listing 6.9). This function handles both device and file access exactly the same way. For every call, `DosRead()` tries to read the number of bytes requested. It will continue to read bytes for every subsequent call until an EOF condition occurs, at which time it returns 0 bytes read.

Listing 6.9   The `DosRead()` function.

```
UCOUNT
DosRead (COUNT hndl, UCOUNT n, BYTE FAR *bp, COUNT FAR *err)
{
    sft FAR *s;
    WORD sys_idx;
    sfttbl FAR *sp;
    UCOUNT ReadCount;

    /* Test that the handle is valid                            */
    if(hndl < 0)
    {
        *err = DE_INVLDHNDL;
        return 0;
    }

    /* Get the SFT block that contains the SFT                  */
    if((s = get_sft(hndl)) == (sft FAR *)-1)
    {
        *err = DE_INVLDHNDL;
        return 0;
    }
```

The algorithm for DosRead() starts in a way similar to other DOS-C system calls. It begins by validating the arguments passed to it. In this case, it verifies that the handle passed by the user is valid or at least within range. Next, it gets the SFT block that corresponds to the handle and checks if it is valid. If not, it returns an error indicating an invalid handle. The final check is for permissions. If the file is not open or does not have write permission, DosRead() exits and returns an error code indicating that this was an invalid file access.

Next it checks structure member sft_flags to see if the sft refers to a device. If the sft does refer to a file, DosRead() proceeds to do a device read. DosRead() first tests for end-of-file and exits immediately if it is. This prevents DosRead() from going any further if the device is no longer open or cannot produce more data. Next it handles raw (binary) and cooked (command line edit) modes. It examines the structure member sft_flags to see if SFT_FBINARY is set. This flag differentiates between the two modes. If it is set, DosRead() performs a direct

---

Listing 6.9   The DosRead( ) function — continued.

```
    /* If not open or write permission - exit              */

    if(s -> sft_count == 0 || (s -> sft_mode & SFT_MWRITE))
    {
        *err = DE_INVLDACC;
        return 0;
    }

    /* Do a device read if device                          */
    if(s -> sft_flags & SFT_FDEVICE)
    {
        request rq;

        /* First test for eof and exit immediately if it is   */
        if(!(s -> sft_flags & SFT_FEOF) || (s -> sft_flags & SFT_FNUL))
        {
            s -> sft_flags &= ~SFT_FEOF;
            *err = SUCCESS;
            return 0;
        }
```

driver call that attempts to read the number of requested bytes. On return
from the device driver call, DosRead() examines the request packet
for errors and invokes the error handler if an error occurred. If no
error occurred, it returns the number of bytes the driver actually read.
DosRead() gets this from the request header structure member
r_count. In this way, the device driver controls the transfer. This is an
excellent way to perform direct reads and writes to a device.

Listing 6.9   *The* DosRead() *function — continued.*

```
        /* Now handle raw and cooked modes                      */

        if(s -> sft_flags & SFT_FBINARY)
        {
            rq.r_length = sizeof(request);
            rq.r_command = C_INPUT;
            rq.r_count = n;
            rq.r_trans = (BYTE FAR *)bp;
            rq.r_status = 0;
            execrh((request FAR *)&rq, s -> sft_dev);
            if(rq.r_status & S_ERROR)
            {
                REG i;
                BYTE buff[FNAME_SIZE+1];

                fbcopy(s -> sft_name, (BYTE FAR *)buff,
                        FNAME_SIZE);
                buff[FNAME_SIZE+1] = 0;
                for(i = FNAME_SIZE; i > 0; i--)
                    if(buff[i] == ' ')
                        buff[i] = 0;
                    else
                        break;
                char_error(&rq, buff);
            }
            else
            {
                *err = SUCCESS;
                return rq.r_count;
            }
        }
```

If SFT_FBINARY is not set, DosRead() proceeds to read from the device in a "cooked" mode, i.e., the characters are processed for break detection, flow control is enabled, etc. There are two possible ways to get the data in this mode. The first is an attempt to read directly from the standard input device. DosRead() examines the sft structure member sft_flags to see if SFT_FSTDIN is set. If it is, it calls sti() to complete

---

Listing 6.9   The DosRead() function — continued.

```
        else if(s -> sft_flags & SFT_FSTDIN)
        {
            if(!check_break())
            {
                kb_buf.kb_size = LINESIZE - 1;
                kb_buf.kb_count = 0;
                sti((keyboard FAR *)&kb_buf);
                fbcopy((BYTE FAR *)kb_buf.kb_buf, bp,
                        kb_buf.kb_count);
                *err = SUCCESS;
                return kb_buf.kb_count;
            }
            else
            {
                *err = SUCCESS;
                return 0;
            }
        }
        else
        {
            if(!check_break())
            {
                *bp = _sti();
                *err = SUCCESS;
                return 1;
            }
            else
            {
                *err = SUCCESS;
                return 0;
            }
        }
    }
```

the read and returns the number of bytes read. If not, it calls the device driver support function and exits with the number of bytes read. The second way to get the data in a "cooked" mode is as an attempt to read from a device driver and _sti() is called instead.

When it is not a device driver, DosOpen() relies on rdwrblock() to perform the file read, returning the count or an error condition if an error occurs. In this context, DosOpen() is simply a pass through to the fs support function call that was covered in detail in Chapter 4.

When you are through with the file, you must close it and return resources back to DOS-C. DosClose() is the DOS personality function that implements the Close system call. It performs in a fashion similar to other calls you have seen.

Listing 6.9   The DosRead() function — continued.

```
    else                          /* a block read              */
    {
        if(!check_break())
        {
            COUNT rc;

            ReadCount = rdwrblock(s -> sft_status, bp, n,
                                  XFR_READ, &rc);
            if(rc != SUCCESS)
            {
                *err = rc;
                return 0;
            }
            else
            {
                *err = SUCCESS;
                return ReadCount;
            }
        }
        else
        {
            *err = SUCCESS;
            return 0;
        }
    }
    *err = SUCCESS;
    return 0;
}
```

As you have seen before, DosClose() (Listing 6.10) first verifies that the handle is valid or within range. Next, it gets the SFT block that corresponds to the handle. If it is not a valid handle, it returns a DOS error indicating that the handle is invalid. In order to close the file, the

---

Listing 6.10   The DosClose() function.

```
COUNT
DosClose (COUNT hndl)
{
    psp FAR *p = MK_FP(cu_psp,0);
    sft FAR *s;

    /* Test that the handle is valid                          */
    if(hndl < 0)
        return DE_INVLDHNDL;

    /* Get the SFT block that contains the SFT                */
    if((s = get_sft(hndl)) == (sft FAR *)-1)
        return DE_INVLDHNDL;

    /* If this is not opened another error*/
    if(s -> sft_count == 0)
        return DE_ACCESS;

    /* now just drop the count if a device, else              */
    /* call file system handler                               */
    if(s -> sft_flags & SFT_FDEVICE)
    {
        p -> ps_filetab[hndl] = 0xff;
        s -> sft_count -= 1;
        return SUCCESS;
    }
    else
    {
        p -> ps_filetab[hndl] = 0xff;
        s -> sft_count -= 1;
        if(s -> sft_count > 0)
            return SUCCESS;
        else
            return dos_close(s -> sft_status);
    }
}
```

final test verifies that the file is open, so that you do not incorrectly return resources that were not allocated, potentially crashing DOS-C. So, if the file is not open, you return an invalid access error.

Now examine whether or not the file is a device. If it is, DosClose() just decrements the SFT count. This allows devices to be opened and shared by duplicating the handle using the Duplicate system call. Otherwise, it calls the file system handler to close the disk file.

This function completes the tour of the DOS personality functions. This gives you a good idea of how to give a system call a particular flavor. I will now look at the other DOS-C personality function set.

FCB Function Support

I will now examine the File Control Block or FCB function calls. These calls provide the old CP/M-style FCB function call personality. Their design is similar to the DOS function calls. Both are layers that perform the direct function of the matching system call. Also, both create a translation service into the three managers: file system, task, and memory.

The FCB function calls also rely heavily on a kernel data structure. The difference is that this structure does not exist solely within the kernel, but is resident in user space. This places some additional restrictions on DOS-C. All designs of FCB support functions must allow for users to modify the FCB or complete an operation without a logical termination, i.e., open a file but never close it.

The FCB is a well-known MS-DOS data structure. It is inherited from its CP/M ancestor and retains many fields that are similar to its CP/M counterpart. It also forms the basis for the DOS-C fcb data structure (Listing 6.11). It is important that this data structure closely matches the MS-DOS version because it is the only means of communicating file-related information for the FCB system calls.

Structure member fcb_drive contains the drive number for the file. By convention, the drives are mapped as 0 = default, 1 = A, etc. The filename is contained in two members: fcb_fname and fcb_fext. These are left-aligned, space-filled representations of the traditional DOS file.ext filename. The next structure member, fcb_cublock, contains the current file position represented as a block number where the block is defined as 128 records. It is mainly used for sequential read/write operations. In order to further refine the current file position, the next structure member, fcb_recsiz, specifies the logical record size in bytes.

Starting with the next member, the fcb starts to record file and directory information. The member fcb_fsize contains the file size in bytes that is initialized when a file is opened or created. DOS-C updates this member with each write operation. File date and time are recorded in structure members fcb_date and fcb_time.

By convention, DOS-C reserves an area of the FCB in order to meet the requirement that users may modify the FCB or complete an operation without a logical termination. The next group of structure members are reserved for DOS-C. The first member in the reserved group is fcb_sftno. This member is used to map the FCB to an internal SFT structure. By starting at fcb_sftno, DOS-C can follow the links down into the f_node structure used by the file system manager. For a device, the next two members, fcb_attrib_hi and fcb_attrib_lo form an attribute mode for files and devices.

Listing 6.11 The *fcb* data structure.

```
/* File Control Block (FCB)                                                  */
typedef struct
{
    BYTE    fcb_drive;              /* Drive number 0=default, 1=A, etc       */
    BYTE    fcb_fname[FNAME_SIZE];  /* File name                              */
    BYTE    fcb_fext[FEXT_SIZE];    /* File name Extension                    */
    UWORD   fcb_cublock;           /* Current block number of                */
                                   /* 128 records/block, for seq. r/w        */
    UWORD   fcb_recsiz;            /* Logical record size in bytes, default = 128 */
    ULONG   fcb_fsize;             /* File size in bytes                      */
    date    fcb_date;              /* Date file created                      */
    time    fcb_time;              /* Time of last write                     */
                                   /* the following are reserved by system   */
    BYTE    fcb_sftno;             /* Device ID                              */
    BYTE    fcb_attrib_hi;         /* share info, dev attrib word hi         */
    BYTE    fcb_attrib_lo;         /* dev attrib word lo, open mode          */
    UWORD   fcb_strtclst;          /* file starting cluster                  */
    UWORD   fcb_dirclst;           /* cluster of the dir entry               */
    UBYTE   fcb_diroff;            /* offset of the dir entry end reserved   */
    UBYTE   fcb_curec;             /* Current block number of                */
    ULONG   fcb_rndm;              /* Current relative record number         */
} fcb;
```

When the device is a file, `fcb_strtclst` records the file's starting cluster and `fcb_dirclst` specifies the cluster that contains the file's directory entry. In order to easily locate the directory entry, structure member `fcb_diroff` contains the offset of the directory entry itself. When combined with `fcb_dirclst`, a directory entry is fully qualified.

By MS-DOS design, the remaining structure entries are user members that are used in random read, write, and seek operations. The first of these members is `fcb_curec`, which contains the current block number. Again to fully specify a record, the structure member `fcb_rndm` contains the current relative record number of the random operation.

As I have done before, I will study the FCB personality functions by studying the steps needed to open, read, and close a file. The first function I will look at is `FcbOpen()`. This function takes a single `far` pointer that points to the FCB in the users memory area as does virtually all of the FCB functions.

`FcbOpen()` (Listing 6.12) begins by attempting to get a free SFT entry. In DOS-C, all calls use an SFT entry to map the FCB in user space into a kernel data structure. Should `FcbOpen()` be unable to get

---

Listing 6.12   The `FcbOpen()` *function.*

```
BOOL FcbOpen(lpXfcb)
xfcb FAR *lpXfcb;
{
    WORD sft_idx;
    sft FAR *sftp;
    struct dhdr FAR *dhp;
    BYTE buff[FNAME_SIZE+FEXT_SIZE + 3];
    fcb FAR *lpFcb;
    COUNT FcbDrive;

    /* get a free system file table entry                        */
    if((sftp = FcbGetFreeSft((WORD FAR *)&sft_idx)) == (sft FAR *)-1)
        return DE_TOOMANY;

    /* Build a traditional DOS file name                         */
    lpFcb = CommonFcbInit(lpXfcb, buff, &FcbDrive);
```

an SFT entry, it returns a DOS error indicating that too many files are open. Next, FcbOpen() converts the XFCB entry (an extended FCB) into an FCB and extracts the filename from the FCB into a local array. It builds a traditional DOS filename in the form of a:file.ext that is understood by the support functions.

Listing 6.12   The FcbOpen() function — continued.

```
/* check for a device                                        */
/* if we have an extension, can't be a device                */
if(IsDevice(buff))
{
    for(dhp = (struct dhdr FAR *)&nul_dev; dhp !=
        (struct dhdr FAR *)-1; dhp = dhp -> dh_next)
    {
        if(FcbFnameMatch((BYTE FAR *)buff, (BYTE FAR *)dhp ->
            dh_name, FNAME_SIZE, FALSE))
        {
            sftp -> sft_count += 1;
            sftp -> sft_mode = O_RDWR;
            sftp -> sft_attrib = 0;
            sftp -> sft_flags =
                (dhp -> dh_attr & ~SFT_MASK) | SFT_FDEVICE |
                  SFT_FEOF;
            sftp -> sft_psp = cu_psp;
            fbcopy(lpFcb -> fcb_fname, sftp -> sft_name,
                    FNAME_SIZE+FEXT_SIZE);
            sftp -> sft_dev = dhp;
            lpFcb -> fcb_sftno = sft_idx;
            lpFcb -> fcb_curec = 0;
            lpFcb -> fcb_recsiz = 0;
            lpFcb -> fcb_fsize = 0;
            lpFcb -> fcb_date = dos_getdate();
            lpFcb -> fcb_time = dos_gettime();
            lpFcb -> fcb_rndm = 0;
            return TRUE;
        }
    }
}
```

In a fashion similar to DosOpen(), FcbOpen() checks for a device because it must initialize the SFT differently for devices than for disk files. If it is not a device, it calls dos_open() to open the disk file. Once this is done, FcbOpen() proceeds to populate the FCB with information necessary to both the user and DOS-C. In essence, FcbOpen() performs many of the same functions as DosOpen(), but functions in a slightly different manner because of the method used to call it.

File reads are handled by FcbRead() (Listing 6.13). As with FcbOpen(), FcbRead() accepts a far pointer into user space that points to the user's FCB. The first step that FcbRead() takes in servicing the

---

Listing 6.12   The FcbOpen() function — continued.

```
    sftp -> sft_status = dos_open(buff, O_RDWR);
    if(sftp -> sft_status >= 0)
    {
        lpFcb -> fcb_drive = FcbDrive;
        lpFcb -> fcb_sftno = sft_idx;
        lpFcb -> fcb_curec = 0;
        lpFcb -> fcb_recsiz = 128;
        lpFcb -> fcb_fsize = dos_getfsize(sftp -> sft_status);
        dos_getftime(sftp -> sft_status,
          (date FAR *)&lpFcb -> fcb_date,
          (time FAR *)&lpFcb -> fcb_time);
        lpFcb -> fcb_rndm = 0;
        sftp -> sft_count += 1;
        sftp -> sft_mode = O_RDWR;
        sftp -> sft_attrib = 0;
        sftp -> sft_flags = 0;
        sftp -> sft_psp = cu_psp;
        fbcopy((BYTE FAR *)&lpFcb -> fcb_fname,
                (BYTE FAR *)&sftp -> sft_name, FNAME_SIZE+FEXT_SIZE);
        return TRUE;
    }
    else
        return FALSE;
}
```

system call is to convert the extended FCB to an FCB if necessary. It then retrieves the SFT block that contains the sft by using the member fcb_sftno. If it cannot retrieve the block, the file was probably not opened and FcbRead() returns an error. If the FCB correctly links to an

---

Listing 6.13   The `FcbRead()` ***function.***

```
BOOL FcbRead(lpXfcb, nErrorCode)
xfcb FAR *lpXfcb;
COUNT *nErrorCode;
{
    sft FAR *s;
    fcb FAR *lpFcb;
    LONG lPosit;
    COUNT nRead;
    psp FAR *p = MK_FP(cu_psp,0);

    /* Convert to fcb if necessary                          */
    lpFcb = ExtFcbToFcb(lpXfcb);

    /* Get the SFT block that contains the SFT              */
    if((s = FcbGetSft(lpFcb -> fcb_sftno)) == (sft FAR *)-1)
        return FALSE;

    /* If this is not opened another error*/
    if(s -> sft_count == 0)
        return FALSE;

    /* Now update the fcb and compute where we need to      */
    /* position to.                                         */
    lPosit = ((lpFcb -> fcb_cublock * 128) + lpFcb -> fcb_curec)
     * lpFcb -> fcb_recsiz;
    if(dos_lseek(s -> sft_status, lPosit, 0) < 0)
    {
        *nErrorCode = FCB_ERR_EOF;
        return FALSE;
    }

    /* Do the read                                          */
    nRead = dos_read(s -> sft_status, p -> ps_dta, lpFcb -> fcb_recsiz);
```

open file, FcbRead() must now update the FCB and compute the correct file position. With this position computed, FcbRead() updates the
file with a call to dos_lseek(), returning an error if it cannot properly
perform the seek.

Listing 6.13   The *FcbRead( ) function — continued.*

```
/* Now find out how we will return and do it.            */
if(nRead == lpFcb -> fcb_recsiz)
{
    *nErrorCode = FCB_SUCCESS;
    FcbNextRecord(lpFcb);
    return TRUE;
}
else if(nRead < 0)
{
    *nErrorCode = FCB_ERR_EOF;
    return TRUE;
}
else if(nRead == 0)
{
    *nErrorCode = FCB_ERR_NODATA;
    return FALSE;
}
else
{
    COUNT nIdx, nCount;
    BYTE FAR *lpDta;

    nCount = lpFcb -> fcb_recsiz - nRead;
    lpDta = (BYTE FAR *)&(p -> ps_dta[nRead]);
    for(nIdx = 0; nIdx < nCount; nIdx++)
        *lpDta++ = 0;
    *nErrorCode = FCB_ERR_EOF;
    FcbNextRecord(lpFcb);
    return FALSE;
}
}
```

Once the file pointer is updated, FcbRead() performs the read operation through a call to dos_read() using the link to the f_node in the sft structure member sft_status, transferring the data to the DTA as indicated in the ps_dta member. The number of bytes to be transferred is the number of bytes in a record for this FCB as stored in fcb structure member fcb_recsiz.

Based on the return, FcbRead() will return in any of three ways. If the number of bytes read matches the record size, the FCB is updated by a call to FcbNextRecord() and returns normally. If it returns fewer bytes read than a record size, FcbRead() returns with an EOF error code. If the number read is exactly zero, FcbRead() returns an error indicating that no data was read.

Finally, the function FcbClose() (Listing 6.14) is called to close the file. Again, a pointer to the user's FCB in user space is passed in and operated on. As you saw with FcbRead(), FcbClose() converts the user's extended FCB to an FCB if necessary. It also gets the SFT block that contains the sft and returns an error if it was not opened.

Next, FcbClose() checks to see if the file opened was a device. If it was, it decrements the sft_count member of the sft and returns. Otherwise, it sets the file's time, date, and size by calling dos_setftime() and dos_setfsize(). With this information updated, FcbClose() calls dos_close() and exits.

Listing 6.14   The *FcbClose( )* function.

```
BOOL FcbClose(lpXfcb)
xfcb FAR *lpXfcb;
{
    sft FAR *s;
    fcb FAR *lpFcb;

    /* Convert to fcb if necessary                            */
    lpFcb = ExtFcbToFcb(lpXfcb);

    /* Get the SFT block that contains the SFT                */
    if((s = FcbGetSft(lpFcb -> fcb_sftno)) == (sft FAR *)-1)
        return FALSE;

    /* If this is not opened another error                    */
    if(s -> sft_count == 0)
        return FALSE;

    /* now just drop the count if a device, else              */
    /* call file system handler                               */
    if(s -> sft_flags & SFT_FDEVICE)
    {
        s -> sft_count -= 1;
        return TRUE;
    }
    else
    {
        s -> sft_count -= 1;
        if(s -> sft_count > 0)
            return SUCCESS;
        else
        {
            /* change time and set file size                  */
            dos_setftime(s -> sft_status,
                        (date FAR *)&lpFcb -> fcb_date,
                        (time FAR *)&lpFcb -> fcb_time);
            dos_setfsize(s -> sft_status, lpFcb -> fcb_fsize);
            return dos_close(s -> sft_status) == SUCCESS;
        }
    }
}
```

## Chapter 7

Command Line Interpreter

Until now, I have been concerned primarily with the details of the DOS-C kernel. I have looked at the system call mechanisms, file system functions, and memory management and task management functions. I have also dug into the device driver interface and investigated the C-to-assembly language interface for calling standard DOS device drivers. With all these details exposed, it's time to look at other components of the operating system, beginning with the user-level commands. A good place to start is with DOS-C's version of command.com.

You may start by asking "How does command.com fit in?" In DOS-C, it is the first program started by the kernel and the parent of every other program started through either a command entered at the keyboard or sets of commands in a batch file. It is the component of the operating system responsible for maintaining the master environment and running the system initialization program, `autoexec.bat`. In many ways, it is similar to the `init` program that runs on many UNIX systems, but it combines the features of command line parsing and initialization. Unlike `init`, `command.com` does not keep a list of programs to monitor or maintain multiple levels that allow single-user and multi-user operation.

Starting the Command Line Interpreter

In Chapter 3, I examined the four stages of a DOS-C boot. In the final stage, the DOS-C kernel loads `command.com`, or any other program specified in the `config.sys` file, and is the first program to run with the DOS-C kernel. It establishes the environment for all subsequent programs executed under DOS-C. What you don't know is the method used to start this program. Take a closer look at how DOS-C starts `command.com`.

Loading `command.com` is a four-step process and is fairly straightforward. The reason is simple: if you allow DOS-C users to replace `command.com` with another command line interpreter of their choice, you need to get that information from the user. There are many ways to do this. You may have a special `init` program that reads a configuration file and starts the command line interpreter. This is the UNIX model where `init` spawns a special set of programs to listen to a serial line or the console. These programs specifically look for a login and then look up the command line interpreter to run from a field contained in the `/etc/passwd` file. You may also build the command line interpreter into the kernel. An operating system that implements this model is CP/M. However, DOS-C is based on the MS-DOS model that specifies the default command line interpreter `command.com` and loads it unless it is directed otherwise by an entry in the file `config.sys`. The DOS-C kernel uses a process broken into three phases to facilitate the `config.sys` switch (Figure 7.1).

The DOS-C kernel uses a data structure, `config`, to hold information regarding the number of buffers, the number of files, and the name of the command line interpreter. This data structure is initialized with default values when the kernel is compiled. The reason you initialize this structure with default values is simple: you need to read a file in order to determine the user kernel configuration. You need to allocate space for file structures and buffers before you can access the file system to read `config.sys`. The kernel uses this initial set to initialize some file system portions of the kernel in preparation of processing `config.sys` in the function `init_kernel()`. It calls the function `PreConfig()` to perform this initial allocation of file system data structures.

With preliminary file system data structures in place, the kernel then processes `config.sys`, replacing the initialized portions of the `config`

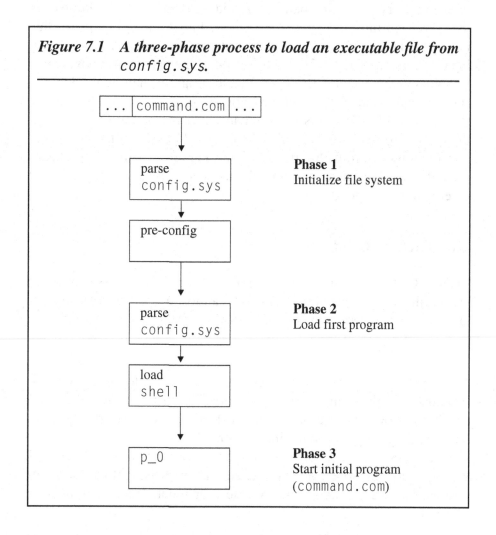

Figure 7.1   *A three-phase process to load an executable file from* `config.sys`.

data structure with those values parsed from the contents of config.sys. One of those structure members is cfgInit. This member contains either the string "command.com" as initialized at compile time or the string parsed from the command= line in config.sys. Also, the remainder of any line parsed from the command= line in config.sys is stored in the structure member cfgInitTail. This allows special parameters such as stack size or environment buffer size to be passed to the start-up command.

When the kernel is ready to execute its first process, it invokes the function p_0(). This function looks up the program it will execute by referring to the command name stored in config.cfgInit and passes the command tail config.cfgInitTail. Function p_0() packs this into a standard DOS exec_blk data structure and calls DosExec(), the internal version of int 21h Fn 00h, to start the program. By using this technique, any standard DOS executable program may be used as the start up command. By convention, you start command.com identically to the way MS-DOS starts its command line interpreter. Once started, the kernel suspends its execution and control passes to the command line interpreter. From this point on, the kernel remains in the background, only to execute system calls as needed by command.com and other application programs.

command.com

In DOS-C, the command line interpreter is command.com. As mentioned earlier, this is by convention and can be almost any other program. For example, during debugging of the DOS-C command.com, the line

```
command=command.exe /p
```

is inserted into config.sys so that the version executed is the one that contains all debugging information normally stripped from the .com file. This allows the use of a standard DOS source-level debugger to simplify the debugging of the interpreter.

The DOS-C version of command.com is written in C. No special coding tricks or foolery are associated with it. It is a simple program designed to accept, parse, and execute user input. It has a number of built-in commands identical to the MS-DOS version and allows the grouping of commands into a single file known as a batch file. Like MS-DOS, DOS-C command.com batch processing allows for some simple program control instructions that test various parameters and modify flow of control based on the results of these tests. It also maintains the global environment block so that programs have access to the global environment variables.

Architecture

The architecture for command.com is simple. It is organized as a table-driven executive with a collection of stand-alone functions used to implement the internal commands. This executive performs a linear search through the table in order to find the function that corresponds to the command. If there is a match, the executive executes the corresponding function; otherwise, it defaults to the terminating entry. This process continues forever, repeating the prompt-command cycle for every user command.

Take a closer look at how command.com works. What you see from the initial examination is that it is very much a standard C program. Like all conventional C programs, execution starts at main() (Listing 7.1).

Listing 7.1   The command.com main() function.

```
VOID main()
{
    COUNT nread;
    BOOL bool_FLAG = FALSE;
    BOOL cflag;
    BYTE FAR *cmd_tail;
    BYTE *p_ptr;
    extern UWORD _psp;
    psp FAR *p;
    COUNT driveno = -1;
    BYTE pattern[MAX_CMDLINE] = "";
    BYTE path[MAX_CMDLINE] = "", esize[MAX_CMDLINE] = "";
```

Listing 7.1   The `command.com main()` *function — continued.*

```
/* Initialize the interpreter                         */
p = MK_FP(_psp, 0);
switchchar = '/';
batch_FLAG = FALSE;
argv[0] = args[0];
argv[1] = (BYTE *)0;
args[0][0] = '\0';
*tail = '\0';
env = (BYTE FAR *)MK_FP(p -> ps_environ, 0);
cmd_tail = MK_FP(_psp, 0x81);
fstrncpy((BYTE FAR *)tail, cmd_tail, 0x7f);

pflag = cflag = FALSE;
dosopt("$d$p*[e:pc]+", (BYTE FAR *)tail,
       &driveno, path, pattern, esize, &pflag, &cflag);

/* Get the passed-in Environment size and make certain we   */
/* allocate enough space                                    */
EnvSize = EnvSizeUp();
if(EnvSize < ENV_DEFAULT)
    EnvSize = ENV_DEFAULT;
if(*esize != '\0')
{
    COUNT size = atoi(esize);

    bool_FLAG = EnvAlloc(size);
    EnvSize = size;
}
else
    bool_FLAG = EnvAlloc(EnvSize);

if(!bool_FLAG)
    error_message(OUT_ENV_SPACE);

/* Check what PROMPT is set in env to over ride default    */
p_ptr = EnvLookup("PROMPT");
if(p_ptr != (BYTE *)0)
    scopy(p_ptr, prompt_string);
else
    scopy(dflt_pr_string, prompt_string);
```

command.com starts by initializing a number of internal variables. It then examines the command tail, searching for options passed when DOS-C invoked command.com. These options modify the execution of the interpreter. For example, the /p option tells command.com that this instance is the root interpreter (since other instances of command.com may be invoked) and must not exit, thereby fixing this copy as the root process for DOS-C. command.com may also be invoked to execute a single command with the /c option, so it must modify its internal code

---

Listing 7.1   The command.com main() function — continued.

```
    /* Check what PATH is set in env to over ride default    */
    p_ptr = EnvLookup("PATH");
    if(p_ptr != (BYTE *)0)
        scopy(p_ptr, path);
    else
        scopy(dflt_path_string, path);
    if(!cflag)
    {
        if(pflag)
        {
            /* Special MS-DOS compatability initialization,    */
            /* all command shells terminate onto themselves,   */
            /* but we always terminate at the root shell.      */
            /* If anyone complains, we'll change it.           */
#ifndef DEBUG
            p -> ps_parent = _psp;
#endif

            /* Try to exec autoexec.bat                        */
            bootup = TRUE;
            *tail = '\0';
            if(!batch(".\\autoexec.bat", tail))
            {
                *tail = '\0';
                cmd_date(1, argv);
                cmd_time(1, argv);
                bootup = FALSE;
            }
        }
```

Listing 7.1 **The** `command.com main()` *function* **—** *continued.*

```
        else
        {
            /* Announce our version                              */
            printf(ANNOUNCE, copyright);
#ifdef SHWR
            printf("**** Shareware version ****\nPlease register
                your copy.\n");
#else
            printf("\n\n");
#endif
        }

        FOREVER
        {
            default_drive = DosGetDrive();
            put_prompt(prompt_string);
            if((nread = DosRead(STDIN, (BYTE FAR *)cmd_line,
                MAX_CMDLINE)) < 0)
                    continue;
            do_command(nread);
        }
    }
    else
    {
        BYTE FAR *p;

        default_drive = DosGetDrive();
        for(p = cmd_tail; *p != '\r'; p++)
        {
            if(*p == '/' && (*(p + 1) == 'c' || *(p + 1) == 'C'))
                break;
        }
        p += 2;
        fstrncpy((BYTE FAR *)cmd_line, p, 0x7f);
        for(nread = 0; *p != '\r'; nread++, p++);
        ++nread;
        do_command(nread);
    }
}
```

path to comply with the user's single-command execution. Other options modify command.com buffer space for environment and stacks. For these reasons, the parsing of command line variables occurs very early on in main( ).

With command line options parsed, main( ) continues to initialize areas that command.com variables affected, such as prompt and path. Once completed, it examines whether or not it is in the single-command mode or the iterative-command interpreter mode. If it is in the single-command mode, the remainder of the command line is copied into a local buffer and main( ) invokes do_command( ) to execute the single command. If it is not executing a single command, command.com assumes it is in the interpreter mode.

The code interpreter mode examines the flags set earlier in main( ). If command.com encountered the /p option, it knows that it was invoked as the root interpreter by the kernel. Within DOS-C, only the root interpreter receives the /p option, so command.com attempts to execute autoexec.bat. Executing this command on start-up is a DOS de facto standard to which all DOS users are well accustomed. Should autoexec.bat not be on the boot disk, command.com gives the user the chance to update the operating system's date and time. This is also a DOS standard and dates back to IBM PC and XT days when there was no system clock, and the start-up of DOS was the only opportunity to set the system time. If command.com did not receive the /p option, the interpreter merely announces its invocation in an attempt to differentiate itself from the root interpreter.

Once command.com completes all this preliminary set-up, it falls into the command loop, an infinite loop that outputs the prompt and then reads from standard input. Once the user enters a command, the buffer that received the keyboard input is passed to do_command( ), which executes the command. Because this loop never exits, the sequence repeats for as long as the interpreter is active. However, command.com can exit, but I will examine this mechanism later.

Both code paths, the single-command path and iterative-command loop, share the function do_command( ) (Listing 7.2). This is the function that actually carries out the execution of individual commands. do_command( ) is responsible for input and output redirection. It also parses the command tail into an argument list for use by internal commands.

Listing 7.2 The *do_command( ) function.*

```
VOID do_command(nread)
COUNT nread;
{
    REG struct table *p;
    REG BYTE *lp;
    COUNT index = 0;
    BYTE Input[MAX_CMDLINE], Output[MAX_CMDLINE];
    BOOL AppendMode;
    COUNT OldStdin = -1, OldStdout = -1, ErrorCode;
    BOOL IORedirected = FALSE;

    if(nread <= 0)
        return;
    cmd_line[nread] = '\0';

    /* Pre-scan the command line and look for any re-directs    */
    *Input = *Output = '\0';
    AppendMode = FALSE;
    Redirect(cmd_line, Input, Output, &AppendMode);
    IORedirected = (*Input != '\0' || *Output != '\0');
    if(*Input != '\0')
    {
        COUNT Handle;

        if(!DosDupHandle(STDIN, (COUNT FAR *)&OldStdin,
           (COUNT FAR *)&ErrorCode))
        {
            RestoreIO(OldStdin, -1);
            error_message(INTERNAL_ERR);
            return;
        }
```

Both the single-command mode and the iterative-command interpreter share a single algorithm. This algorithm is encompassed by do_command(). When initially invoked, do_command() prescans the command line for redirection. This is done through the invocation of Redirect(), which scans the command line looking for all entries that

---

Listing 7.2   The *do_command( ) function — continued.*

```
        Handle = DosOpen((BYTE FAR *)Input, O_RDWR);
        if((Handle < 0) || (!DosForceDupHandle(Handle, STDIN,
                           (COUNT FAR *)&ErrorCode)))
        {
            RestoreIO(OldStdin, -1);
            error_message(INTERNAL_ERR);
            return;
        }
        DosClose(Handle);
    }

    if(*Output != '\0')
    {
        COUNT Handle;

        if(!DosDupHandle(STDOUT, (COUNT FAR *)&OldStdout,
                         (COUNT FAR *)&ErrorCode))
        {
            RestoreIO(-1, OldStdout);
            error_message(INTERNAL_ERR);
            return;
        }

        if(AppendMode)
        {
            if((Handle = DosOpen((BYTE FAR *)Output, O_RDWR)) < 0)
            {
                RestoreIO(-1, OldStdout);
                error_message(INTERNAL_ERR);
                return;
            }
            DosSeek(Handle, 2, 0l);
        }
        else
            Handle = DosCreat((BYTE FAR *)Output,
                              D_NORMAL | D_ARCHIVE);
```

are preceded by "<" or ">", which are conventionally used for I/O redirection. Once this scan is complete, do_command() proceeds to change the standard input and output file descriptors with the DOSAPI calls DosDupHandle(), DosOpen(), and DosForceDupHandle(). This has the net effect of saving the original file descriptor, so you can recover it later, and redirecting the input or output stream. This is done once for the input stream and again for the output stream.

With I/O redirection out of the way, do_command() moves on to parsing the command line for possible use by internal commands. The internal commands use an internal representation of argc and argv similar to that in any conventional C program. Once the argument list is built, do_command() moves on to execute the command. It does this by first looking for the special case of a simple drive change. This is a two-character string pointed to by argv[0], where the second character is

---

Listing 7.2   The *do_command( ) function — continued.*

```
        if((Handle < 0) || (!DosForceDupHandle(Handle, STDOUT,
                            (COUNT FAR *)&ErrorCode)))
        {
            RestoreIO(-1, OldStdout);
            error_message(INTERNAL_ERR);
            return;
        }
        DosClose(Handle);
    }

    for(argc = 0; argc < 16; argc++)
    {
        argv[argc] = (BYTE *)0;
        args[argc][0] = '\0';
    }
    lp = scanspl(cmd_line, args[0], '/');

    if(args[0][0] == '@')
    {
        at_FLAG = TRUE;
        index++;
    }
    else
        at_FLAG = FALSE;
```

Listing 7.2 **The** do_command() **function — continued.**

```
    /* If preceeded by a @, swallow it, it was taken care of    */
    /* elsewhere.  Also, change case so that our command verb   */
    /* is case sensitive.                                       */
    while(args[0][index] != '\0')
    {

        if(at_FLAG)
            args[0][index-1] = tolower(args[0][index]);
        else
            args[0][index] = tolower(args[0][index]);
        index++;
    }
    if(at_FLAG)
        args[0][index-1] = '\0';

    argv[0] = args[0];
    /* this kludge is for an MS-DOS wart emulation              */
    tail = skipwh(lp);

    for(argc = 1; argc < NPARAMS; argc++)
    {
        lp = scan(lp, args[argc]);
        if(*args[argc] == '\0')
            break;
        else
            argv[argc] = args[argc];
    }

    if(*argv[0] != '\0')
    {
        /* Look for just a drive change command, and execute    */
        /* it if found.                                         */
        if(argv[0][1] == ':' && argv[0][2] == NULL)
        {
            BYTE c = argv[0][0];

            if(c >= 'a' && c <= 'z')
                c = c - 'a' + 'A';
            if(c >= 'A' && c <= 'Z')
                DosSetDrive(c - 'A');
        }
```

":". Should this be the case, the current drive is changed through a DOSAPI call to `DosSetDrive()`, and `do_command()` exits. If it is not a drive change, `do_command()` checks for a special request to the Help subsystem. If `do_command()` identifies it as being such a request, it builds a Help command line, invokes `help.exe` and exits.

Finally, if the command is neither a drive change nor a call to the Help subsystem, `do_command()` attempts to invoke either an internal or external command to satisfy the user request. It does this through a call to `lookup()` (Listing 7.3), which performs the actual command lookup. The internal command's corresponding function is then executed by invoking the returned pointer to the internal function. Once the command completes, `do_command()` restores the I/O to its original state and exits.

Listing 7.2   The do_command() *function — continued.*

```
        /* It may be a help command request.                  */
        else if((argv[1][0] == switchchar) && (argv[1][1] == '?'))
        {
            strcpy(tail, " ");
            strcat(tail, argv[0]);
            strcat(tail, "\r\n");
            argc = 2;
            argv[1] = argv[0];
            argv[0] = "help";
            argv[2] = 0;
            ExecCmd(argc, argv);
            if(IORedirected)
                RestoreIO(OldStdin, OldStdout);
        }
        /* do a normal command execution                      */
        else
        {
            p = lookup(commands, argv[0]);
            (*(p -> func))(argc, argv);
            if(IORedirected)
                RestoreIO(OldStdin, OldStdout);
        }
    }
}
```

When the default entry is invoked, it attempts to take the command parsed from the command line and locate an external program to fulfill the user's request. It follows a predetermined sequence of searches based on filename extensions. This is conventionally $*$.com, $*$.exe, and $*$.bat, in that order. It spawns the external program, if found, and the external command returns to this point when it terminates.

DOSLIB

Before I proceed much further, I need to explain what the DOSAPI that I mentioned previously is all about. What you have seen is a sequence of calls such as DosOpen(), DosSetDrive(), etc. MS-DOS shows the influences of early operating system designs in the assembly language system call, wherein MS-DOS, and in turn DOS-C, implements system calls as software interrupts, and system call parameters are passed in the processor register set. The choice of registers seems arbitrary and does not follow any high-level language conventions. Consequently, special arrangements must be made for high-level languages such as C, in effect hiding the assembly language details from the C program.

Listing 7.3   The `lookup()` *function.*

```
struct table *lookup(p, token)
struct table *p;
BYTE *token;
{
    while(*(p -> entry) != '\0')
    {
        if(strcmp(p -> entry, token) == 0)
            break;
        else
            ++p;
    }
    return p;
}
```

There are many ways to hide the system call details from the C code. It is possible to embed the system calls directly in a C function through the use of parameterized C macros. This technique is used by operating systems such as Linux. It is also possible to implement a message-passing scheme where the system call parameters are placed into a message packet and sent to the kernel. This is the Minix technique. What I have done in DOS-C is define a logical set of C functions that act as the DOSAPI. These functions are then specially constructed to accept normal C parameters and convert them to the `int 21h` system calls accepted by DOS-C. In doing this, I have defined the DOS-C DOSAPI.

Examine the `DosOpen()` API call (Listing 7.4) to study the anatomy of an API call. The call starts out as a normal C function defining two parameters, `FileName`, normally passed as a `far` pointer in `ds:dx`, and `FileAccess`, normally in `ah`. The function returns a 16-bit integer that represents either the handle of the opened file or an error code if negative.

On starting the function `DosOpen()`, the parameter `FileName` is broken into segment and offset components suitable for the `int 21h` call. Although you could have deferred this until later, after you make the transition to assembly language, the `FP_SEG()` and `FP_OFF()` macros are used instead. This way, you don't need to second guess how the compiler encodes `far` pointers. `DosOpen()` then shifts gears and drops into assembly language using the C directive `asm`. Once in assembly language, the parameters are recovered from the function invocation and `moved` into the correct registers. `DosOpen()` then loads the function number into the `al` register and proceeds to make an `int 21h` call. When the `int 21h` call returns, the return values are checked for errors and the return value is set accordingly.

At this point you may be wondering why the API calls are coded in this fashion. After all, you could code the API function entirely in assembly language or use the more common `int21()` or `intdos()` functions. Any of these choices would work, but the tradeoffs are apparent. If you code the API in assembly language, you must have intimate knowledge of how the compiler passes parameters in function calls, and the use of `int21()` and related calls runs the risk of using

self-modifying code or code built on the stack. Although you don't necessarily want to put this code in read-only memory, the code should not
preclude this option. The hybrid approach chosen here represents the
best compromise between the two and is an approach typically seen in
many operating system APIs.

Listing 7.4   *The DOSAPI function* `DosOpen()`.

```
COUNT DosOpen(FileName, FileAccess)
COUNT FileAccess;
BYTE FAR *FileName;
{
    UWORD  FileName_seg = FP_SEG(FileName);
    UWORD  FileName_off = FP_OFF(FileName);
    COUNT  Handle;

    asm {
    push   ds
    push   cx
    mov    cx,FileName_seg
    mov    dx,FileName_off
    mov    al,byte ptr FileAccess
    mov    ds,cx

    mov    ah,0x3d
    int    0x21

    pop    cx
    pop    ds

    mov    Handle,ax

    jnc    out
    neg    ax
    mov    Handle,ax

    }
out: return Handle;
}
```

Internal Commands

I will now look at two internal commands, cd and exit. These two functions are representative of internal commands in command.com. In the case of cd, the design is virtually identical to all the other internal commands. exit, however, is a special case that is used to quit the interpreter. Because of this, exit needs to do some special housekeeping before exiting. Start by examining cd.

The function cd() (Listing 7.5) starts out by initializing the automatic variable OldDrive. You do this to store the drive that is logged in when the flow of execution entered the cd command, so you can return to the original logged-in drive in the event that you encounter an error anywhere within the function. Next, you do a command line sanity check as a simple method of checking the syntax. This sanity check

---

Listing 7.5   The cd() function.

```
BOOL cd(argc, argv)
WORD argc;
BYTE *argv[];
{
    COUNT OldDrive, NewDrive = -1;
    BYTE CurDir[MAX_CMDLINE] = "";

    /* Initialize where we are                                      */
    OldDrive = DosGetDrive();

    /* Do command line sanity checks                                */
    if(argc > 2)
    {
        error_message(INV_NUM_PARAMS);
        return FALSE;
    }

    /* get command line options and switch to the requested drive */
    dosopt("$d*", (BYTE FAR *)tail, &NewDrive, CurDir);
    if(NewDrive < 0)
        NewDrive = default_drive;
    DosSetDrive(NewDrive);
```

is to make certain that the user only enters commands of the form cd or cd d: (or other drive designation). Should a syntax error occur, an error is returned to the main command loop.

With the syntax check out of the way, it's time to parse the command line for options. In the case of cd(), you are looking for an optional directory parameter, which may contain an optional drive specification that you need to separate. cd() uses dosopt() to search for an optional drive, and if it encounters one, it calls DosSetDrive() to temporarily switch to that drive. If no directory is present in the second argument, cd simply echoes the current parameter. If one is present, cd() proceeds to change the directory, and then switches back to the original drive, completing the cd() operation.

Listing 7.5   The cd() function — continued.

```
    /* Do pwd function for no parameter case              */
    if(*CurDir == '\0')
    {
        DosPwd(NewDrive + 1, CurDir);
        printf("%c:\\%s\n\n", 'A' + NewDrive, CurDir);
        DosSetDrive(OldDrive);
        return TRUE;
    }


    /* Otherwise, change the directory, and then switch back    */
    /* to the old directory.                                    */
    if((DosCd((BYTE FAR *)CurDir)) != SUCCESS)
    {
        error_message(INV_DIR);
        DosSetDrive(OldDrive);
        return FALSE;
    }
    else
    {
        DosSetDrive(OldDrive);
        return TRUE;
    }
}
```

cd illustrates the general design of every internal command. First, it performs a syntax check. Next, it proceeds to extract the information it needs by parsing the arguments that were passed on invocation. Finally, the command is implemented using DOSAPI calls. Error checks are performed at each step and a means of recovery is designed into each step. When the command exits because of an error or another reason, it returns an exit code that indicates either success (TRUE) or failure (FALSE). In the event of a failure, the function calls an error routine to deliver an error message. All commands within command.com follow this design.

cmd_exit() (Listing 7.6) handles the built-in exit command. The design of this function deviates significantly from other built-in commands because of its special nature. Within DOS, the exit command is reserved to terminate any invocation of the command interpreter except for the initial one started when DOS starts up. DOS-C follows this same convention and supplies an exit command similar to that found in any other DOS, including MS-DOS. The one deviation that DOS-C makes is that the kernel invokes the initial command.com with a /p switch to indicate that it is the permanent shell.

When you examine cmd_exit(), you find that it begins with code that examines whether or not it is contained within a permanent shell. This early check aborts cmd_exit() before proceeding into the body of the function, guaranteeing that this invocation of command.com doesn't exit.

Another feature of command.com is that it can return values that convey information, such as error codes back to the program that invoked the command. This feature is typically seen in batch files, allowing the batch file to return a code indicating either an error condition caused the termination of the batch program or successful completion. A return code of zero means, by convention, success. command.com also returns success unless otherwise instructed by an argument to the exit command. cmd_exit() does this by a simple examination of argc. If there is only a single argument, it exits with a success code; otherwise, it converts the first argument to a numeric value and returns that number as the return code.

External Commands

ExecCmd() is the central point where command.com invokes external commands. This is an important function since the vast majority of DOS commands and applications are invoked through this interface.

Listing 7.6   The cmd_exit() function.

```
BOOL cmd_exit(argc, argv)
COUNT argc;
BYTE FAR *argv[];
{
#ifndef DEBUG
    /* Don't exit from a permanent shell                          */
    if(pflag)
        return TRUE;
#endif

    /* If no values passed, return errorvalue = 0                 */
    if(argc == 1)
        DosExit(0);

    /* otherwise return what the user asked for                   */
    else
    {
        COUNT ret_val;
        static BYTE nums[] = "0123456789";
        BYTE FAR *p;

        for(ret_val = 0, p = argv[1]; isdigit(*p); p++)
        {
            COUNT j;

            for(j = 0; j < 10; j++)
                if(nums[j] == *p)
                    break;
            ret_val += j;
        }
        DosExit(ret_val);
    }
    return TRUE;
}
```

Listing 7.7 illustrates the function ExecCmd(), which seems excessive at first glance. You shouldn't jump to conclusions too early in the game, however, because starting a program and accurately keeping track of its environment can consume significant resources.

Start by looking at the overall design. All commands are entered through the command line interface you examined earlier. ExecCmd() parses the command line and breaks it into an internal argc/argv structure similar to the C calling convention for main(). It then takes the first argument and attempts to match it against the internal dispatch table.

Listing 7.7   The ExecCmd() function.

```
BOOL ExecCmd(argc, argv)
COUNT argc;
BYTE *argv[];
{
    exec_blk exb;
    COUNT err;
    BYTE tmppath[64];
    COUNT idx;
    BOOL ext = FALSE;
    BYTE *extp;
    COUNT len;
    BYTE *lp;
    CommandTail CmdTail;
    fcb fcb1, fcb2;
    static BYTE *extns[2] =
    {
        ".com",
        ".exe"
    };
    static BYTE *batfile = ".bat";
    BYTE PathString[MAX_CMDLINE];
    BYTE Path[MAX_CMDLINE], *pPath;

    /* Build the path string and create the full string that    */
    /* includes ".\" so that the current directory is searched   */
    /* first.  Note that Path is initialized outside the loop.   */
    strcpy(Path, ".\\");
    strcpy(PathString, EnvLookup("PATH"));
    pPath = PathString;
```

Listing 7.7   The `ExecCmd()` *function — continued.*

```
do
{
    /* Build a path to the command.                        */
    if(*pPath == ';')
        ++pPath;
    strcpy(tmppath, Path);
    if(*tmppath != '\0' && !((tmppath[strlen(tmppath) - 1] !=
                        '\\') == 0))
        strcat(tmppath, "\\");
    strcat(tmppath, argv[0]);

    /* batch processing                                    */
    /* search for an extension in the specification        */
    for(idx = len = strlen(argv[0]) ; idx > 0 &&
        idx > (len - FEXT_SIZE - 2); --idx)
    {
        if(argv[0][idx] == '.')
        {
            ext = TRUE;
            extp = &argv[0][idx];
            break;
        }
    }

    /* If no extension was found, the entire path was      */
    /* specified and we do not append an extension.        */
    if(!ext)
    {
        strcat(tmppath, batfile);
        extp = batfile;
    }

    /* if it ends with a '.bat' (either user supplied or   */
    /* previously added), try to run as a batch.           */
    if((strcmp(extp, batfile) == 0) && batch(tmppath, tail))
    {
        if(pflag && bootup)
            bootup = FALSE;
        return TRUE;
    }
```

The algorithm used for the match stops when a terminating null string entry is encountered. The accompanying entry points to `ExecCmd()` make it the default function to be invoked when looking for internal commands. In this fashion, an external command is treated no differently than an internal command. By making it the default entry, you have guaranteed that `command.com` searches the internal commands before loading an external command.

Listing 7.7   The `ExecCmd()` function — continued.

```
        /* tail comes in as a string with trailing newline.   */
        /* Convert it to a return only and put it into CmdTail */
        /* format                                              */
        CmdTail.ctCount = (argc > 1) ? strlen(tail) : 1;
        strcpy(CmdTail.ctBuffer, " ");
        strcpy(&CmdTail.ctBuffer[1], (argc > 1) ? tail : "");
        CmdTail.ctBuffer[CmdTail.ctCount] = '\0';
        if(CmdTail.ctCount < LINESIZE)
            CmdTail.ctBuffer[CmdTail.ctCount] = '\0';
        rtn_errlvl = 0;
        exb.exec.env_seg = FP_SEG(env);
        exb.exec.cmd_line = (CommandTail FAR *)&CmdTail;

#if PARSEFN
        if(argc > 1)
        {
            DosParseFilename((BYTE FAR *)argv[1],
                        (fcb FAR *)&fcb1, 0);
            exb.exec.fcb_1 = (fcb FAR *)&fcb1;
        }
        else
            exb.exec.fcb_1 = (fcb FAR *)0;
        if(argc > 2)
        {
            exb.exec.fcb_2 = (fcb FAR *)&fcb2;
            DosParseFilename((BYTE FAR *)argv[2],
                        (fcb FAR *)&fcb2, 0);
        }
        else
            exb.exec.fcb_2 = (fcb FAR *)0;
```

Listing 7.7   The `ExecCmd()` ***function — continued.***

```
#else
        exb.exec.fcb_1 = (fcb FAR *)0;
        exb.exec.fcb_2 = (fcb FAR *)0;
#endif

        for(idx = 0; idx < 2; idx++)
        {
            strcpy(tmppath, Path);
            if(*tmppath != '\0' &&
                !((tmppath[strlen(tmppath) - 1] != '\\') == 0))
                strcat(tmppath, "\\");
            strcat(tmppath, argv[0]);
            if(!ext)
            {
                strcat(tmppath, extns[idx]);
                extp = extns[idx];
            }
            if(!(strcmp(extp, extns[idx]) == 0))
                continue;
            if((rtn_errlvl = err = DosExec((BYTE FAR *)tmppath,
                                        (exec_blk FAR *)&exb))
                                        != SUCCESS)
            {
                switch(err)
                {
                case DE_FILENOTFND:
                    continue;

                case DE_INVLDFUNC:
                    rtn_errlvl = INV_FUNCTION_PARAM;
                    goto errmsg;

                case DE_PATHNOTFND:
                    rtn_errlvl = PATH_NOT_FOUND;
                    goto errmsg;

                case DE_TOOMANY:
                    rtn_errlvl = TOO_FILES_OPEN;
                    goto errmsg;
```

Listing 7.7   The `ExecCmd()` **function — continued.**

```
                    case DE_ACCESS:
                        rtn_errlvl = ACCESS_DENIED;
                        goto errmsg;

                    case DE_NOMEM:
                        rtn_errlvl = INSUFF_MEM;
                        goto errmsg;

                    default:
                        rtn_errlvl = EXEC_ERR;
                    errmsg:
                        error_message(rtn_errlvl);
                        return FALSE;
                }
            }
            else
            {
                rtn_errlvl = DosRtnValue() & 0xff;
                return TRUE;
            }
        }
        if(err < 0 || idx == 2)
        {
            if(!(err == DE_FILENOTFND || idx == 2))
            {
                error_message(EXEC_FAIL);
                return FALSE;
            }
            continue;
        }
    }
    while(*Path = '\0', pPath = scanspl(pPath, Path, ';'),
        *Path != '\0');
    error_message(BAD_CMD_FILE_NAME);
    return FALSE;
}
```

Once `ExecCmd()` is invoked, it initializes and searches for a command in a fixed manner that includes searching for `.bat`, `.com`, and `.exe` forms of the command. This search is performed along fixed directories as specified in the `path` environment variable. Once `ExecCmd()` finds the command, it creates the proper environment before executing the external command. When the external command returns, `ExecCmd()` checks for errors before exiting to the interpreter's main loop, which adds to the robustness of the design.

Take a closer look at the algorithm behind `ExecCmd()`. `ExecCmd()` starts by initializing the internal execution path variable. It builds the path string and creates a path that begins with ".\" so that the current directory is searched first. Next, `ExecCmd()` uses a C `do ... while` construct to repeat the search along each directory contained within the string constructed earlier. For each iteration, `ExecCmd()` builds a `.bat` form of the command and searches for it by invoking the function `batch()` (Listing 7.8). If it is a batch file, `ExecCmd()` returns; otherwise, it tries to load a binary executable.

To execute a binary (`.com` or `.exe`) file, `ExecCmd()` uses an `exec` block data structure for use in an `int 21h fn 4bh`. It starts by creating a command tail for use in the system call. It also initializes the `exec` block with the master environment segment. Next, it initializes the two `fcb` entries by parsing the command line using the Parse Filename system call (`int 21h fn 29h`). Finally, it attempts to execute a `.com` version then an `.exe` version of the command through the `DosExec()` API call. The order of `.com` then `.exe` is controlled by order of entry in the `extns` table.

Listing 7.8   The `batch()` function.

```
BOOL batch(file)
BYTE *file;
{
    COUNT idx;
    COUNT file_parse();
    BOOL parse();
```

Listing 7.8   The `batch()` function — continued.

```c
    /* check to see if currently processing a batch file       */
    /* if so clean up                                          */
    if(batch_FLAG)
    {
        COUNT fp;

        if((fp = DosOpen((BYTE FAR *)file, O_RDONLY)) < 0)
            return FALSE;
        else
            DosClose(fp);
        batch_FLAG = FALSE;
        for(idx = 0; idx < NPARAMS; idx++)
            posparam[idx][0] = '\0';
        for(label_cnt = 0; label_cnt < MAX_LABELS; label_cnt++)
        {
            labels[label_cnt].lb_name[0] = '\0';
            labels[label_cnt].lb_posit = '\0';
        }
        DosClose(fileptr);
    }

    /* open batch file for reading                             */
    if((fileptr = DosOpen((BYTE FAR *)file, O_RDONLY)) < 0)
    {
        batch_FLAG = FALSE;
        return FALSE;
    }

    /* Ok, now set mode to batch and initialize positional     */
    /* parameter array.                                        */
    echo_FLAG = TRUE;
    batch_FLAG = TRUE;
    shift_offset = 0;
    for(idx = 0; idx < NPARAMS; idx++)
        strcpy(posparam[idx], args[idx]);

    /* deal with command line                                  */
    file_parse(fileptr);
    return TRUE;
}
```

On return from the `DosExec()` API call, `ExecCmd()` examines the return code for error conditions. It does this for two reasons. First, it needs to see if it found the command. `DosExec()` uses this return information to proceed to the next entry, if necessary. It does this with a simple C `continue` statement, bypassing all other error tests. Second, it needs to translate the error codes into useful user feedback in the form of error messages. These messages are an absolute necessity for the user. Without them, `command.com` would be nearly useless.

Batch Commands

Batch commands execute in a fashion similar to external commands but with significant differences. Unlike a binary file, `command.com` executes these commands by sequentially reading a text file that contains a sequence of commands. It reads the file one line at a time, interpreting each line.

Command line interpreters in many operating systems, such as UNIX and its derivatives, possess similar features. Unfortunately, the DOS batch command language it is not nearly as well-structured as these other interpreters. The batch language that DOS uses tends to be very simple with not much support for flow control or structured programming. In fact, some language features such as the `call` batch command did not occur until late in the development of MS-DOS. Without the `call` batch command, you cannot nest batch commands simply, and the lack of this command was a problem. Many users found batch programming without the `call` command extremely restrictive, and it took quite a bit of demand from the user community before Microsoft responded.

The restrictive DOS batch language is somewhat unfortunate for the DOS-C design. One of our design goals is that DOS-C be functionally equivalent to MS-DOS. Maintaining compatibility with the standard DOS batch language is necessary to prevent breaking applications that rely on it. There is quite a bit of risk associated with deviating from the standard because many batch files in the DOS software pool may be affected by the changes. For example, these batch files are used in various applications as control programs coordinating multiple subcommands, in installation procedures as installation scripts, and by many users for local customization. The `command.com` batch language is also the script language used by the system start-up command `autoexec.bat`. As a result, the implementation of the `command.com` batch scripting language should match the MS-DOS version. This makes the design somewhat cumbersome in spots. In fact, in the process of matching the original MS-DOS design, you quickly understand how unstructured assembly language coding led to the unstructured batch language design.

Look at the DOS-C method used to execute a batch command. Batch commands are initiated in `command.com` through the invocation of the `DosExec()` function. `ExecCmd()` attempts to execute batch commands before binary commands and uses a call to the function `batch()` to execute the batch command. If the command is not a batch command, `batch()` returns a `FALSE`, allowing `ExecCmd()` to continue. However, if it is a batch command, `batch()` executes the batch file by reading it line by line and executing each line.

The function `batch()` takes only a single argument, a pointer to a string representing the batch command to be executed. Armed with the filename, `batch()` attempts to open the batch file, but only if it is not currently processing one, as indicated by the global variable `batch_FLAG`. If it is in the middle of a batch file, `batch()` first checks to see if the file can be opened. If it is not opened, it returns, thereby allowing for error processing. If the file is good, it's opened and internal variables are reset. At this point you may wonder why you need all these steps before ever getting to the meat of batch processing. The answer is simple, if not obscure: the code is here to properly handle the chaining of batch files, a `command.com` feature heavily relied on. It is the first of many special code snippets necessary for correct MS-DOS emulation.

In any event, with the possible exception of a bad batch file specification, `batch()` proceeds to open the new batch file for input, again keeping track of failures to allow for error processing. It then initializes a number of internal variables and begins parsing the file for commands through a call to `file_parse()` (Listing 7.9), returning `TRUE` as an indicator for success on completion.

Listing 7.9   The `file_parse()` ***function.***

```
COUNT file_parse(fileptr)
WORD fileptr;
{
    WORD rc_FLAG;
    COUNT nread;
    COUNT pass;
    BYTE line[MAX_CMDLINE];
    BYTE *p;

    for(pass = 1; pass <= NUM_PASSES; pass++)
    {
        default_drive = DosGetDrive();
        do
        {
            BOOL eof = FALSE, eol = FALSE;

            for(*line = 0, p = line, nread = 0;
                !eol && !eof && (nread < MAX_CMDLINE); )
            {
                if(DosRead(fileptr, (BYTE FAR *)p, 1) != 1)
                {
                    eof = TRUE;
                    break;
                }
                switch(*p)
                {
                case CTL_Z:
                case '\r':
                    *p = '\0';
                    eol = TRUE;
                    break;

                case '\n':
                    *p = '\0';
                    continue;

                default:
                    ++nread;
                    ++p;
                    break;
                }
            }
        }
```

Listing 7.9 **The** `file_parse()` **function — continued.**

```
            expand(cmd_line, line);

            default_drive = DosGetDrive();
            p = skipwh(cmd_line);

            /* dummy nread to get by a blank line              */
            if(pass > 1 && *p == '\0')
            {
                nread = 1;
                if(echo_FLAG)
                {
                    if(eof)
                        printf("\n");
                    put_prompt(prompt_string);
                    printf("\n");
                }
                if(!eof)
                    continue;
                else
                    break;
            }

            if(echo_FLAG && pass > 1 && *p != '@' && *p != ':' &&
               nread > 0)
            {
                printf("\n");
                put_prompt(prompt_string);
                printf("%s\n", cmd_line);
            }

            if((rc_FLAG = parse(cmd_line, strlen(cmd_line), pass))
               != TRUE)
            {
                printf("ERROR parsing batch file\n");
                return (rc_FLAG);
            }
        } while (nread > 0);
        DosSeek(fileptr, 0, 0L);
    }

    DosClose(fileptr);
    batch_FLAG = FALSE;
    return TRUE;
}
```

The `file_parse()` function is heart of batch processing. This function implements the basic structure of the batch command language. `file_parse()` builds a symbol table to keep track of labels and is responsible for expanding batch variables. Take a closer look at how this is done.

The basic control structure of `file_parse()` is a `for` loop that loops through the body of the `file_parse()` code. `file_parse()` loops through the batch file twice. The first time, it goes through and builds a label table that contains an entry for each label plus the file offset for that label. This allows the flow of control functions, such as `goto`, to quickly position to the correct offset in the file. In each pass, `file_parse()` first builds a line from the file, terminating on either end-of-file, carriage return, or line feed. Once the line is read, substitutions are made for each `command.com` variable, such as %0 through %9 and other shell variables. It does this by a call to the function `expand()`. Once the line is expanded, it echoes the line if needed, then executes the command by a call to `parse()` (Listing 7.10).

The function `parse()` is the final link in batch processing. `parse()` modifies its behavior based on the `pass` variable. If `pass` is equal to 1, a label, signified as a string preceded with a ":", is entered into the symbol

---

Listing 7.10   The `parse()` function.

```
BOOL parse(line, nread, pass)
BYTE line[];
COUNT nread;
COUNT pass;
{
    cursor = skipwh(line);

    if(pass == 1)
    {
        if(*cursor == ':')
            return label_bat(++cursor);
        else
            return TRUE;
    }
```

table by a call to `label_bat()`. All other commands are bypassed. For all other `pass` values, the label lines are bypassed, but command lines are executed by calling `do_command()`, which is the same function that parses each individual command line while in the interactive mode.

As you can see, a batch file is handled nearly identically to individual commands typed in by the user. The only exception is that you keep track of labels. What you don't see is how flow of control is executed. If you look at the dispatch table, you see that you have built-in commands for `if` and `goto`. In the DOS-C version of `command.com`, each statement that controls flow is a built-in command similar to the `dir` command. Look at the `goto` command to understand how these commands work.

All flow of control statements are similar in design. Like `goto_bat()` (Listing 7.11), they initially examine the `batch_FLAG` and return immediately if `command.com` is not in batch mode. This flag disables these commands during interactive operation and is set only when

---

Listing 7.10   The `parse()` function — continued.

```
    switch(*cursor) {
        case '@':
            at_FLAG = TRUE;
            cursor = skipwh(++cursor);
            do_command(nread);
            at_FLAG = FALSE;
            break;
        case ':':
            /* labels processed in pass 1              */
            break;
        default:
            do_command(nread);
            break;
    }
return TRUE;
}
```

command.com is in batch mode. You need to disable flow of control statements during interactive operation since you cannot reposition the input file (the keyboard in this case).

When a flow of control statement needs to transfer the command flow to a new location denoted by a label, it searches the symbol table for the label. When it finds the label, the function retrieves the corresponding file offset and uses it to reposition the file, effectively performing a jump to the statement following the label. For conditional commands, the evaluation of the condition prior to performing the lookup modifies the batch file execution, but the code is virtually identical. Although not complicated, the method is very effective.

Listing 7.11   The `goto_bat()` *function.*

```
BOOL goto_bat()
{
    COUNT label_cnt = 0;
    BYTE *lp;

    if(!batch_FLAG)
        return FALSE;

    lp = skipwh(tail);

    while(labels[label_cnt].lb_name != '\0')
    {
        if(strcmp(labels[label_cnt].lb_name, lp) == 0)
        {
            DosSeek(fileptr, 0, (LONG)labels[label_cnt].lb_posit);
            return TRUE;
        }
        label_cnt++;
    }
    error_message(LABEL_NOT_FOUND);
    return FALSE;
}
```

Other Options for Command Interpreters

Other command line interpreters are available for MS-DOS, and each has its advantages and disadvantages. However, this chapter demonstrates the fundamentals of operating system command line interpreters. In general, these techniques are similar to those used in other command line interpreters found in other operating systems. Although the system calls may be different, the general principles are the same.

The DOS-C version of command.com is not a full implementation for MS-DOS. Two features are noticeably missing. First, the MS-DOS command.com makes a special int 28h call that is used to multiplex print.com while idle. MS-DOS needs this because of its single thread of execution. It must cooperatively share its thread of control with print.com. Also missing are pipes. MS-DOS does not support interprocess communications, again because of its single thread of execution. Pipes are implemented as temporary files in MS-DOS that are deleted after being read. This is a simple but effective technique that simulates pipes well.

These are features that can be easily added to our command.com, if so desired. As with all of DOS-C, the source is yours to enhance and change as you please.

## Chapter 8

DOS-C Kernel: Putting It All Together

If you have read the book to this chapter, you now have an understanding of the design behind DOS-C. You have gone over the high-level design and taken different tours through the kernel to examine various aspects of the kernel. You have studied some of the fundamental kernel data structures and discovered how these data structures relate to both DOS-C and MS-DOS. But you haven't built DOS-C yet.

This chapter covers the actual building of the kernel and debugging techniques. The concept of building a program and debugging it with the use of remote debuggers may be new to many readers, so I will go through a step-by-step build and test session of the kernel.

Organizing the Project

Any software project usually has some structure associated with it. In the smallest of projects, all the necessary files are usually kept in a single directory. In larger projects, individual components are kept in separate directories and built in one common area. DOS-C follows the latter model.

At the highest level, the project is organized into five directories: lib, doc, hdr, dist, and src (Figure 8.1). Each of these directories holds a major component of DOS-C. The src directory is further divided into subdirectories: ipl, boot, command, drivers, fs, kernel, misc, utils, and tmp. This organization adds structure to the project. Is every project organized this way? No, not necessarily, and this directory structure may not make sense for another project. You may not even organize your own operating system project this way, but the object is to organize your files in a way that makes the overall project easy to maintain.

At the top level, the dist directory is an abbreviation for distribution and holds all files necessary for a binary distribution. It is essentially an image of a bootable floppy that you would give to someone to

---

Figure 8.1   DOS-C directory structure.

```
+---lib
+---doc
+---hdr
+---dist
+---src
    +---ipl
    +---boot
    +---command
    +---drivers
    +---fs
    +---kernel
    +---misc
    +---utils
    +---tmp
```

try DOS-C. In this directory are `boot.bin`, `ipl.sys`, `kernel.exe`, `command.com`, other executable files, and documentation, including release notes. Had this been a larger project, this directory may have been further subdivided. For example, it may have been divided into one directory for each disk of a distribution. Another possible organization is to use a single directory but have other areas for distribution on different media. Whatever makes the most sense for your project should guide your implementation. For DOS-C, the binaries are distributed two ways: a compressed archive and a snuggle floppy. For this reason, it makes sense that all the files are collected into a single directory, and distribution is built from that directory.

The `doc` directory is where all project documentation goes. This directory includes informational files, release notes, to-do lists, and any other project-related documentation. Had DOS-C been bigger, the directory may have been further divided into separate directories. DOS-C is an active project, and as it grows, so will the documentation. This directory structure may change later, but for now, a single directory with one file per documentation item is all that is necessary.

The next three directories are dedicated to building the executables. They are `lib`, `hdr`, and `src`, abbreviations for libraries, headers, and source, respectively. The first of these directories, `lib`, is a common directory that holds libraries built in the project. The next directory, `hdr`, holds all header files used in the project. The final directory, `src`, is divided into subdirectories, each dedicated to a component of the project.

Again you may wonder about the structure. Simply, you need to make certain that all key components, such as headers and libraries, are kept in areas that are easily accessible from the lower source levels. The general rule for DOS-C is that any project-specific header or library is always two directories up from the one you are in when you build the file. This simple organization simplifies the developer's task and eliminates duplication of files and just plain losing files. Again, this structure may not be ideal for every project, but any project that is large enough should have a file structure. This way, all members of a project team know where all the files are located.

When you organize your project, examine the project. Try to organize the directory structure in a way that makes sense. In the case of DOS-C, the directories are organized by functionality, but for your project you may want to divide files into one directory per executable. It all depends on what makes sense for your particular development environment.

You will also need to consider your build tools. Some make programs include concepts such as view pathing. This is a technique where multiple directories specified in a makefile variable are searched for dependencies, much the same way that command.com searches multiple directories in its path variable. In this case, you may want to create a phantom directory structure for work in progress and use the view path to pick up files from the source tree that are needed for the build but use local files that may be under development.

Building the DOS-C Kernel

The build organization of DOS-C is a separate makefile for each major component. Each makefile is named, with a .mak extension, after the component it builds. For example, the makefile for kernel.exe is kernel.mak (Listing 8.1) and for ipl.sys is ipl.mak. This allows for rapid identification and association of the component and the makefile. There is no overall makefile. To build the entire package, invoke the batch file build.bat that resides in the top directory.

Listing 8.1   Kernel makefile kernel.mak.

```
#
# Makefile for Borland C++ 3.1 for kernel.exe
#
# $Header:   D:/dos-c/src/kernel/kernel.mav   1.0   02 Jul 1995  8:30:22   patv  $
#
# $Log:   D:/dos-c/src/kernel/kernel.mav   $
#
#   Rev 1.0   02 Jul 1995  8:30:22   patv
#Initial revision.
#

RELEASE = 1.00

.AUTODEPEND
```

Each makefile has, at a minimum, four targets. These targets are used by the top build file and also standardize the operations across each DOS-C component. These targets, `all`, `production`, `populate`, `clobber`, and `clean`, define common makefile operations that you can perform for each DOS-C component. The target `all` builds every

---

Listing 8.1   Kernel makefile `kernel.mak` — ***continued.***

```
#
# Compiler and Options for Borland C++
# ------------------------------------
CC = bcc +kernel.cfg
ASM = TASM
LIB = TLIB
LINK = TLINK
LIBPATH = .
INCLUDEPATH = ..\HDR
CFLAGS      = -v -X -I. -D__STDC__=0 -DDEBUG -DKERNEL -DI86 -DPROTO -DSHWR -DASMSUPT
AFLAGS      = /Mx/Zi/DSTANDALONE=1
LIBS        =..\..\LIB\DEVICE.LIB ..\..\LIB\LIBM.LIB

# where to copy source from
FSSRC = \
  ..\fs\fatfs.c \
  ..\fs\fatdir.c \
  ..\fs\fattab.c \
  ..\fs\dosfns.c \
  ..\fs\fcbfns.c \
  ..\fs\error.c
SUPTSRC = \
  ..\fs\prf.c \
  ..\fs\misc.c \
  ..\fs\dosnames.c \
  ..\fs\syspack.c
IOSRC = \
  ..\fs\blockio.c \
  ..\fs\chario.c

# what to delete when cleaning
COPIEDSRCA = \
  fatfs.c \
  fatdir.c \
  fattab.c \
  dosfns.c \
  fcbfns.c
COPIEDSRCB = \
  error.c \
  prf.c \
  misc.c \
  dosnames.c \
  syspack.c \
  blockio.c \
  chario.c
```

executable for that component. It may consist of one or more .bin,
.com or .exe files. The target populate copies all source files into the
directory. DOS-C uses this target to copy files from fs, the file system
manager directory, into ipl and kernel before building the file. This

---

Listing 8.1   Kernel makefile kernel.mak — *continued.*

```
#               *Implicit Rules*
.c.obj:
  $(CC) -c {$< }

.cpp.obj:
  $(CC) -c {$< }
#               *List Macros*


EXE_dependencies = \
 kernel.obj \
 blockio.obj \
 chario.obj \
 dosfns.obj \
 dsk.obj \
 error.obj \
 fatdir.obj \
 fatfs.obj \
 fattab.obj \
 fcbfns.obj \
 initoem.obj \
 inthndlr.obj \
 ioctl.obj \
 main.obj \
 memmgr.obj \
 misc.obj \
 dosnames.obj \
 prf.obj \
 strings.obj \
 sysclk.obj \
 syscon.obj \
 syspack.obj \
 systime.obj \
 task.obj \
 apisupt.obj \
 asmsupt.obj \
 execrh.obj \
 procsupt.obj \
 stacks.obj

 #        *Explicit Rules*
 all:      production

production: populate kernel.exe
            tdstrip kernel.exe
            copy kernel.exe ..\..\dist
            del *.obj
            del kernel.exe
```

Listing 8.1   Kernel makefile `kernel.mak` **— continued.**

```
populate:  $(FSSRC) $(SUPTSRC) $(IOSRC)
           release $(RELEASE)
           ..\utils\pop $(FSSRC)
           ..\utils\pop $(SUPTSRC)
           ..\utils\pop $(IOSRC)

clobber:   clean
           del kernel.exe

clean:
           ..\utils\rmfiles $(COPIEDSRCA)
           ..\utils\rmfiles $(COPIEDSRCB)
           del *.obj
           del *.bak
           del *.crf
           del *.xrf
           del *.map
           del *.lst
           del *.las

kernel.exe: kernel.cfg $(EXE_dependencies)
  $(LINK) /m/v/c/P-/L$(LIBPATH) @&&|
kernel.obj+
blockio.obj+
chario.obj+
dosfns.obj+
dsk.obj+
error.obj+
fatdir.obj+
fatfs.obj+
fattab.obj+
fcbfns.obj+
initoem.obj+
inthndlr.obj+
ioctl.obj+
main.obj+
memmgr.obj+
misc.obj+
dosnames.obj+
prf.obj+
strings.obj+
sysclk.obj+
syscon.obj+
syspack.obj+
systime.obj+
task.obj+
apisupt.obj+
asmsupt.obj+
execrh.obj+
procsupt.obj+
stacks.obj
kernel          # exe file
kernel          # map file
$(LIBS)
|
```

way, only one copy of each source file exists, and software maintenance is simplified. Both targets `clean` and `clobber` clean the directory of files that are left over after a build occurs. The only difference is that `clean` leaves the product files, whereas `clobber` leaves only the source files that belong in the directory. The target `production` creates the actual version that appears in the `dist` directory. This version is stripped of all debugging information and built without debugging options.

Listing 8.1   Kernel makefile `kernel.mak` ***— continued.***

```
#           *Individual File Dependencies*
kernel.obj: kernel.cfg kernel.asm
     $(ASM) $(AFLAGS) KERNEL.ASM,KERNEL.OBJ

blockio.obj: kernel.cfg blockio.c

chario.obj: kernel.cfg chario.c

dosfns.obj: kernel.cfg dosfns.c

dsk.obj: kernel.cfg dsk.c

error.obj: kernel.cfg error.c

fatdir.obj: kernel.cfg fatdir.c

fatfs.obj: kernel.cfg fatfs.c

fattab.obj: kernel.cfg fattab.c

fcbfns.obj: kernel.cfg fcbfns.c

initoem.obj: kernel.cfg initoem.c

inthndlr.obj: kernel.cfg inthndlr.c

ioctl.obj: kernel.cfg ioctl.c

main.obj: kernel.cfg main.c

memmgr.obj: kernel.cfg memmgr.c

misc.obj: kernel.cfg misc.c

dosnames.obj: kernel.cfg dosnames.c

prf.obj: kernel.cfg prf.c

strings.obj: kernel.cfg strings.c

sysclk.obj: kernel.cfg sysclk.c
```

Now build DOS-C. First, change directory to the DOS-C root directory. Next, enter the command `build` and the process begins. The batch file changes directory to each component directory and executes `make` with the target `production` specified. This usually takes a few minutes. When it completes, the `dist` directory contains all the executable components of DOS-C. This is a simple procedure to build an entire operating system.

Listing 8.1   Kernel makefile `kernel.mak` — continued.

```
syscon.obj: kernel.cfg syscon.c

syspack.obj: kernel.cfg syspack.c

systime.obj: kernel.cfg systime.c

task.obj: kernel.cfg task.c

apisupt.obj: kernel.cfg apisupt.asm
     $(ASM) $(AFLAGS) APISUPT.ASM,APISUPT.OBJ

asmsupt.obj: kernel.cfg asmsupt.asm
     $(ASM) $(AFLAGS) ASMSUPT.ASM,ASMSUPT.OBJ

execrh.obj: kernel.cfg execrh.asm
     $(ASM) $(AFLAGS) EXECRH.ASM,EXECRH.OBJ

procsupt.obj: kernel.cfg procsupt.asm
     $(ASM) $(AFLAGS) PROCSUPT.ASM,PROCSUPT.OBJ

stacks.obj: kernel.cfg stacks.asm
     $(ASM) $(AFLAGS) STACKS.ASM,STACKS.OBJ


proto.h:   $(CSRC) $(LIBCSRC1) $(LIBCSRC2) $(LIBCSRC3) $(LIBCSRC4)
           echo /* proto.h generated by make */ > proto.h
           echo #undef _P >> proto.h
           ..\utils\proto $(CSRC)
           ..\utils\proto $(LIBCSRC1)
           ..\utils\proto $(LIBCSRC2)
           ..\utils\proto $(LIBCSRC3)
           ..\utils\proto $(LIBCSRC4)
           echo Don't forget to edit proto.h for version control

#          *Compiler Configuration File*
kernel.cfg: kernel.mak
  copy &&|
$(CFLAGS)
| kernel.cfg
```

Now, look at another potential activity. Add a feature to the DOS-C kernel. Add function 80h, which outputs a message when invoked. You will need to modify the file inthndlr.c to add the new system call and a test program, test80.c (Listing 8.2), to test the new feature.

The new function call is simple. Use the built-in printf() function to output the test message and the contents of the al register. Then negate the al register and return. It easily could have been a more complex call, similar to some of the other DOS-C system calls, but this serves the purpose.

Edit inthndlr.c (Listing 6.4) and add the code from Listing 8.3 to the int21_service() switch. Build the new kernel in the kernel directory with the command make -f kernel kernel.exe to build a debug version of the kernel. The make builds the kernel and when it's complete, you proceed to make a test disk.

Listing 8.2   Test program `test80.c.`

```
#include <stdio.h>
#include <dos.h>

#define TESTFN 0x80

void main(void)
{
    union REGS regs;

    printf("Start test --\n");
    regs.h.ah = TESTFN;      /* our new system call            */
    regs.h.al = 0x12;
    int86(0x21, &regs, &regs);
    printf("Done.\n Function returned 0x%02x\n", regs.h.al);
}
```

Listing 8.3   New system call 80h.

```
case 0x80:
    printf("System Call received 0x%02x\n", r -> AL);
    r -> AL = -(r -> AL);
    break;
```

First, switch to the `dist` directory and use the `sys` command to create a bootable floppy disk. Next, switch back to the `kernel` directory and copy `kernel.exe` onto the disk, overwriting the distribution kernel. Finally, copy `test80.exe` to the disk. It is that simple to create a test disk. In fact, once you have a bootable floppy disk, all you need to do is copy `kernel.exe` to it because the loader, `ipl.sys`, has the capability to load EXE files. Reboot the system and run `test80`. It returns the message "System Call received 0xee".

Testing the Kernel (or Did I Really Want to Do This?)

Now comes the fun part — testing your new kernel. Many times, you will make a change and the kernel will simply not work. The question is: what do you do then? Take a look at debugging techniques that may be used to debug any stand-alone operating system or embedded application

You have a number of debugging options. The first is to place special messages throughout code. This technique proves quite useful and is quick, low cost, and effective when the program that you are debugging is running, just incorrectly.

With DOS-C, this option is built into the kernel at a few key places. Also, if you look at `int` 21h function 33h subfunction `0fdh` and `0ffh`, you will see two debug toggles used for read and write system calls and system calls in general. These prove quite helpful in determining what a user program is doing.

However, this type of debugging generally cannot handle tougher problems. For example, wild pointers and uninitialized variables require guess work to specifically dump the affected area. Often, this type of problem doesn't manifest itself until a particular application or sequence of events occur. That is when guess work and iterative tests become time consuming.

However, there are other software alternatives. The first is simulation. Using this technique, you build a special version of the program and run it under a well-behaved operating system. DOS-C has a special version called kdb. kdb is a small command line interpreter that simulates system calls into the kernel and allows the user to set and examine variables. It is particularly useful during the initial debugging stages when algorithms are being tested and if a good source-level debugger is available for the particular C compiler you are using.

Another especially useful technique is remote debugging. With this technique, you first load a small debugger onto the target system or in ROM. Then, using the host, load the program to be debugged onto the target system. The host system then allows you to examine levels, single step, set break points, and perform other general source-level debugging. Much of DOS-C application testing was done using a commercially available remote debugger. The package used during DOS-C development required building a special target program that included some special startup code. To handle this, a special makefile was created that built the special debug target. However, these systems cost many hundreds of dollars.

The disadvantage of this technique is that when your target system crashes, you may have no way of performing a post mortem, especially if the code you are testing is in RAM. A wild pointer or an infinite loop can quickly overwrite the program or the data area with usually no way of stopping or breaking the program.

The optimum technique is to use an in-circuit emulator, which has all the advantages of remote debugging, but with additional hardware support. With an in-circuit emulator, you remove the processor from the target under test and replace it with an umbilical to special hardware connected to the host. This hardware simulates the processor and adds special hardware break points not normally available to remote software debuggers. For example, if you have a particular variable that mysteriously changes during debugging, you can set a hardware break point on a write to that one location or range of locations. When a write occurs, the in-circuit emulator will stop and show you exactly which instruction in your code is doing the errant write.

In-circuit emulators also have other features, such as trace stacks, which store some past sequence of instructions. You can turn on the trace buffer and stop, then look at all the history within the buffer. This is especially useful when examining loop terminations and hardware interrupt timing. Some models even have built-in profiling logic that will do a profile of your code nonintrusively.

So, what's the catch? Cost. In-circuit emulators typically cost thousands of dollars and some are literally full-blown workstations that run into the tens of thousands. The price for debugging ease may be expensive.

It is all a matter of your application and budget. If you are writing the operating system for a multimillion dollar realtime application, multiple in-circuit emulators are probably appropriate. If you are doing a midnight engineering job on a budget, remote debugging or simple `printf()` statements will do fine. It is up to you to determine your debugging technique.

Where Do You Go From Here?

It is all up to you. I have presented you with a look into an operating system that performs like a popular well-known operating system, MS-DOS. What you do with it is your choice.

You can use it to customize your environment and build special features into it to support your application. You can also make some minor changes and embed it into an application as the embedded operating system, or you can simply change `main()` and fall directly into your application. Or you can use it as an example to better understand the principles of an operating system.

The choice is yours. You now have a good understanding of DOS-C and the source code is for you to use within the DOS-C license terms. Use it as a tool and good luck.

A Note about Portability

Portability is an issue from which hardly anyone can escape. Developers encounter pressures daily, from management and customers alike, to develop portable code. In a typical project, it may be desirable to use the same source code for different platforms in order to create a wider market for the end product. Also, with technology rapidly changing, the code developed today will probably be reused on the next generation of hardware, necessitating portable design techniques. Additionally, sharing the code base among different processors greatly eases software maintenance problems.

The C programming language is a major contributor to portability. It is ubiquitous and standard. By using C, today's developer accomplishes a significant degree of portability but is also lulled into a false sense of security by the language's features. C library functions may behave differently with different operating systems. There are also semantic problems with the language itself. For example, integer size is variable, depending on the target machine. With segmented architecture processors, another variable is pointer size and memory organization when the pointer is in memory. Yet another difference is the order of bits in bit fields. These, and other factors, present quite a challenge to the developer.

When I originally began the design of the predecessor to DOS-C, I had already tackled portability problems in my designs of real-time and embedded systems using both 8-bit and 16-bit processors. I had also standardized on C in the early 1980s to help bring these products to market faster. This experience proved invaluable when a potential customer approached me to make it available on a non-Intel platform. Encouraged by this customer, I undertook the task of writing a portable operating system.

One important part of the design was to ensure processor independence in the code. The first step in achieving this goal was to minimize the amount of assembly code. The second step was to guarantee portable C code with stringent design rules to guarantee word size and byte order independence.

I guaranteed meeting these design rules with the architecture described in Chapter 3 and by designing a portable set of types that I could change to match the capability of the target processor with the use of the C preprocessor #ifdef. These portable types allow me to guarantee that the physical disk structures are processor neutral. The need for portable types becomes evident by studying the DOS file system. A quick examination of the MS-DOS FAT file system reveals three simple types: 8-bit, 16-bit, and 32-bit. I achieved part of the goal by simply casting the 8-bit type to char and the 32-bit type to long. However, an int is not a good choice for the 16-bit type because its length may be either 16-bit or 32-bit depending on the target processor.

Casting the 16-bit type to a short guarantees the 16-bit length. Additionally, I needed to guarantee the representation of these variables as both signed and unsigned values, so creating another unsigned set guaranteed non-negative values for ANSI-C compilers. Unfortunately, one of the target machines only had K&R C available, which complicated the issue because K&R C doesn't have some unsigned types.

I did not want to depend on familiarity with C types in order to recognize the size of the variable, so I decided to create aliases so they stand out in even the most cursory browsing of the code. I chose the following:

```
typedef char            BYTE;
typedef short           WORD;
typedef long            DWORD;

typedef unsigned char   UBYTE;
typedef unsigned short  UWORD;

typedef unsigned long   ULONG;
typedef short           SHORT;
#ifdef STRICT
typedef signed long     LONG;
#else
#define LONG            long
#endif
```

Pointer size and organization are another area of concern for compatibility. As you know, the 80x86 family has a real mode designed to be compatible throughout the entire family. This guarantees upward compatibility from the 8088. DOS is a real mode operating system originally designed for the 8088. Many of the DOS system calls require far pointers in the form of segment:offset. To take advantage of the 32-bit linear protected mode, a layered product, such as a DOS extender or the Windows GUI, supplies protected mode enhancements.

This means that DOS clone designers must make use of C constructs such as `far` and memory models such as small. Because DOS-C itself is small enough to fit into the small model (separate code and data areas, each less than 64Kb), I decided to take advantage of the small model so that I would have tighter and faster code than if I used the large model (separate code and data areas, each with as much as 1Mb of memory). However, the tradeoff for this decision is that any reference to memory outside of DOS-C must be a `far` reference, and the segment:offset pair must be adjusted. Unfortunately, compilers for linear machines may not support the `far` and `near` keywords. The method I chose to simplify this requirement is:

```
#ifdef <segmented machine>
#  define FAR far
#  define NEAR near
#endif

#ifdef <linear machine>
#  define FAR
#  define NEAR
#endif
```

With this approach you compile all files with a command line option to switch between the architectures (i.e., `-DI86` for 80x86, `-DMC68K` for 680x0, etc.). When the preprocessor completes its pass, architecture-dependent features such as the `far` are expanded to either "far" or " " depending on the architecture.

Finally, the issue of byte order differences between in-memory and disk images of file system variables is addressed whenever data is transferred between a disk data structure member and a memory structure member. The in-memory version is always the native representation by default. I did not want to waste time by manipulating it every time the program needed to access an in-memory variable. However, when I transfer the variable to or from a disk buffer, I invoke the appropriate conversion function. This function is either a preprocessor macro or a C function, depending on whether or not the target is a native 80x86 processor. When I compile a native version, a preprocessor switch bypasses the conversion function for the sake of efficiency. This was handled by the following code:

```
#ifdef NATIVE
#  define getlong(vp, lp) (*(LONG *)(lp)=*(LONG *)(vp))
#  define getword(vp, wp) (*(WORD *)(wp)=*(WORD *)(vp))
#  define getbyte(vp, bp) (*(BYTE *)(bp)=*(BYTE *)(vp))
#  define fgetlong(vp, lp) (*(LONG FAR *)(lp)=*(LONG FAR *)(vp))
#  define fgetword(vp, wp) (*(WORD FAR *)(wp)=*(WORD FAR *)(vp))
#  define fgetbyte(vp, bp) (*(BYTE FAR *)(bp)=*(BYTE FAR *)(vp))
#  define fputlong(lp, vp) (*(LONG FAR *)(vp)=*(LONG FAR *)(lp))
#  define fputword(wp, vp) (*(WORD FAR *)(vp)=*(WORD FAR *)(wp))
#  define fputbyte(bp, vp) (*(BYTE FAR *)(vp)=*(BYTE FAR *)(bp))
#else
VOID getword(VOID *, WORD *);
VOID getbyte(VOID *, BYTE *);
VOID fgetlong(VOID FAR *, LONG FAR *);
VOID fgetword(VOID FAR *, WORD FAR *);
VOID fgetbyte(VOID FAR *, BYTE FAR *);
VOID fputlong(LONG FAR *, VOID FAR *);
VOID fputword(WORD FAR *, VOID FAR *);
VOID fputbyte(BYTE FAR *, VOID FAR *);
#endif
```

The conversion functions themselves are in the file `syspack.c` (Listing A.1).

Listing A.1   `syspack.c`

```
/******************************************************************/
/*                                                              */
/* SYSPACK.C                                                    */
/*                                                              */
/* System Disk Byte Order Packing Functions                    */
/*                                                              */
/* Copyright (c) 1995                                          */
/* Pasquale J. Villani                                         */
/* All Rights Reserved                                         */
/*                                                              */
/* This file is part of DOS-C.                                 */
/*                                                              */
/* DOS-C is free software; you can redistribute it and/or      */
/* modify it under the terms of the GNU General Public License */
/* as published by the Free Software Foundation; either version*/
/* 2, or (at your option) any later version.                   */
/*                                                              */
/* DOS-C is distributed in the hope that it will be useful, but*/
/* WITHOUT ANY WARRANTY; without even the implied warranty of  */
/* MERCHANTABILITY or FITNESS FOR A PARTICULAR PURPOSE.  See   */
/* the GNU General Public License for more details.            */
/*                                                              */
/* You should have received a copy of the GNU General Public   */
/* License along with DOS-C; see the file COPYING.  If not,    */
/* write to the Free Software Foundation, 675 Mass Ave,        */
/* Cambridge, MA 02139, USA.                                   */
/*                                                              */
/******************************************************************/

#include "../../hdr/portab.h"
#include "globals.h"

/* $Logfile:   D:/dos-c/src/fs/syspack.c_v  $ */
#ifndef IPL
static BYTE *syspackRcsId = "$Header:
                        D:/dos-c/src/fs/syspack.c_v   1.3   29
May 1996 21:15:12   patv  $";
#endif
```

Listing A.1   `syspack.c` *— continued*

```
/*
 * $Log:   D:/dos-c/src/fs/syspack.c_v  $
 *
 *    Rev 1.3   29 May 1996 21:15:12   patv
 * bug fixes for v0.91a
 *
 *    Rev 1.2   01 Sep 1995 17:48:42   patv
 * First GPL release.
 *
 *    Rev 1.1   30 Jul 1995 20:50:26   patv
 * Eliminated version strings in ipl
 *
 *    Rev 1.0   02 Jul 1995  8:05:34   patv
 * Initial revision.
 */

#ifdef NONNATIVE
VOID
getlong (REG VOID *vp, LONG *lp)
{
 *lp = (((BYTE *)vp)[0] & 0xff) +
  ((((BYTE *)vp)[1] & 0xff) << 8) +
  ((((BYTE *)vp)[2] & 0xff) << 16) +
  ((((BYTE *)vp)[3] & 0xff) << 24);
}

VOID
getword (REG VOID *vp, WORD *wp)
{
 *wp = (((BYTE *)vp)[0] & 0xff) + ((((BYTE *)vp)[1] & 0xff) << 8);
}

VOID
getbyte (VOID *vp, BYTE *bp)
{
 *bp = *((BYTE *)vp);
}
```

Listing A.1   `syspack.c` — *continued*

```
VOID
fgetword (REG VOID FAR *vp, WORD FAR *wp)
{
 *wp = (((BYTE FAR *)vp)[0] & 0xff) +
  ((((BYTE FAR *)vp)[1] & 0xff) << 8);
}

VOID
fgetlong (REG VOID FAR *vp, LONG FAR *lp)
{
 *lp = (((BYTE *)vp)[0] & 0xff) +
  ((((BYTE *)vp)[1] & 0xff) << 8) +
  ((((BYTE *)vp)[2] & 0xff) << 16) +
  ((((BYTE *)vp)[3] & 0xff) << 24);
}

VOID
fgetbyte (VOID FAR *vp, BYTE FAR *bp)
{
 *bp = *((BYTE FAR *)vp);
}

VOID
fputlong (LONG FAR *lp, VOID FAR *vp)
{
 REG BYTE FAR *bp = (BYTE FAR *)vp;

 bp[0] = *lp & 0xff;
 bp[1] = (*lp >> 8) & 0xff;
 bp[2] = (*lp >> 16) & 0xff;
 bp[3] = (*lp >> 24) & 0xff;
}

VOID
fputword (WORD FAR *wp, VOID FAR *vp)
{
 REG BYTE FAR *bp = (BYTE FAR *)vp;

 bp[0] = *wp & 0xff;
 bp[1] = (*wp >> 8) & 0xff;
}
```

Listing A.1   `syspack.c` *— continued*

```
VOID
fputbyte (BYTE FAR *bp, VOID FAR *vp)
{
 *(BYTE FAR *)vp = *bp;
}
#endif

VOID
getdirent (BYTE FAR *vp, struct dirent FAR *dp)
{
 fbcopy(&vp[DIR_NAME], dp -> dir_name, FNAME_SIZE);
 fbcopy(&vp[DIR_EXT], dp -> dir_ext, FEXT_SIZE);
 fgetbyte(&vp[DIR_ATTRIB], (BYTE FAR *)&dp -> dir_attrib);
 fgetword(&vp[DIR_TIME], (WORD FAR *)&dp -> dir_time);
 fgetword(&vp[DIR_DATE], (WORD FAR *)&dp -> dir_date);
 fgetword(&vp[DIR_START], (WORD FAR *)&dp -> dir_start);
 fgetlong(&vp[DIR_SIZE], (LONG FAR *)&dp -> dir_size);
}

VOID
putdirent (struct dirent FAR *dp, BYTE FAR *vp)
{
 REG COUNT i;
 REG BYTE FAR *p;

 fbcopy(dp -> dir_name, &vp[DIR_NAME], FNAME_SIZE);
 fbcopy(dp -> dir_ext, &vp[DIR_EXT], FEXT_SIZE);
 fputbyte((BYTE FAR *)&dp -> dir_attrib, &vp[DIR_ATTRIB]);
 fputword((WORD FAR *)&dp -> dir_time, &vp[DIR_TIME]);
 fputword((WORD FAR *)&dp -> dir_date, &vp[DIR_DATE]);
 fputword((WORD FAR *)&dp -> dir_start, &vp[DIR_START]);
 fputlong((LONG FAR *)&dp -> dir_size, &vp[DIR_SIZE]);
 for(i = 0, p = (BYTE FAR *)&vp[DIR_RESERVED]; i < 10; i++)
  *p++ = NULL;
}
```

The FreeDOS Project: Official FAQ File

This document is intended to answer frequently asked questions that arise regarding the FreeDOS Project and its goals. Please take the time to read this document to determine if your questions are answered here before contacting project coordinators.

Revision History

3 August 1996, M. "Hannibal" Toal, First Draft

General Information

What Is the Goal of the FreeDOS Project?

The goal of the FreeDOS Project is to create a completely free MS-DOS-compatible operating system. FreeDOS will run on all DOS-capable platforms, from XTs to Pentium Pros.

What Is the Reason for FreeDOS?

There are many users who either do not have access to hardware that is capable of running today's 32-bit operating systems (such as Windows 95, OS/2, and UNIX) or who do not require the complexity of those systems.

FreeDOS will be beneficial for businesses, schools, and organizations because there are no royalty payments. FreeDOS will be an option for hobbyists, hackers, and anyone who would enjoy a chance to examine or customize the source code and inner workings of a real operating system for educational, practical, or recreational purposes.

What Will FreeDOS Not Be?

FreeDOS will not be multitasking. It will not be object oriented. It will not include a Graphical User Interface, a flat memory model, or operate in 32-bit protected mode.

It's not that we do not desire any of these things. It is fun and interesting to speculate about the capabilities of an advanced, 32-bit version of FreeDOS. However, it is the goal of FreeDOS to complete a 16-bit MS-DOS-compatible operating system first, before proceeding in different directions.

Who Owns FreeDOS?

Everybody and nobody. FreeDOS and its associated programs and documentation are protected under the terms of the GNU General Public License (GPL). This means that FreeDOS can be freely copied, so long as there is no attempt to restrict further copying. All copyrights are retained by the original authors. All source code will be publically available. See the full text of the GPL, included with FreeDOS, for more information.

How Is FreeDOS Project Development Organized?

The FreeDOS Project Committee, a group of the senior contributors and participants involved with FreeDOS, is a group of individuals that exists to make decisions and set standards for the FreeDOS Project.

The FreeDOS project consists of a number of specific groups and each of these groups has a coordinator. These groups are the utility, shell, kernel, testing, and documentation groups.

Technical Information

What Exactly Is Included in FreeDOS?

FreeDOS will be a complete, independent operating system. It includes a kernel, a shell, and a full complement of utilities. FreeDOS uses a freeware C compiler for all development work. Each is described separately in the following text.

What Is the DOS-C Kernel?

The *kernel* is the heart of an operating system. It is a special program that contains low-level functions and routines to bridge the gap between hardware and software. Pat Villani's DOS-C is the kernel that forms the basis for the FreeDOS operating system. DOS-C, as the name implies, is written entirely in C and will eventually be 100 percent MS-DOS compatible.

What Is the FreeDOS Shell?

A *shell* is an interactive program that is used to load and run other programs and usually also provides a basic set of tools and functions necessary to make a computer actually useful. Users familiar with MS-DOS will know this program as `command.com`.

The FreeDOS `command.com` (written by Tim Norman in cooperation with several other developers) provides piping, redirection, command line history and editing, several built-in functions, and batch file capability. Future versions will add enhancements such as command aliasing.

What Are the FreeDOS Utilities?

The *utilities* that accompany an operating system are designed to enable a user to perform productive tasks and to control and customize the computing environment.

The FreeDOS utilities include the traditional assortment of programs that MS-DOS users will find familiar, from `attrib` to `vol`. Some will be slightly different, some will be enhanced, and some will be new programs that should have been included with MS-DOS long ago.

What Is the DDS MICRO-C Compiler?

The MICRO-C compiler is a free C compiler writtten by Dave Dunfield of Dunfield Development Systems. MICRO-C is not officially a part of FreeDOS, as it is not published under the terms of the GNU General Public License. MICRO-C is a capable and well-designed free product that was chosen as the standard compiler for FreeDOS programs until a free compiler capable of producing 8086-compatible code is available.

Will FreeDOS Run on Some Particular Machine?

Probably. The DOS-C kernel will run on virtually all hardware that is truly IBM-PC compatible. Very old hardware, or in some cases very new or unusual hardware, may have problems running DOS-C. For example, the DOS-C kernel does not currently deal properly with second hard drives.

Will FreeDOS Run Some Particular Program?

Maybe. Most of the problems that people have when running FreeDOS are due to either bugs or unfinished parts of the DOS-C kernel. The latest version (v0.91a) of the DOS-C kernel is able to run WordPerfect 5.1 and DOOM. Microsoft Windows does not currently run under DOS-C.

Will FreeDOS Be Compatible with Some Particular Version of MS-DOS?

FreeDOS will be compatible with MS-DOS v3.30. The reason I chose version 3.30 as a target is that there are very few fundamental changes between MS-DOS v3.30 and subsequent versions.

Specific Answers to Common Questions

Why Not Use DJGPP?

DJGPP is a fine product, but it generates code for only 386+ machines. Because the goal of FreeDOS is to run on all PC-compatible machines, DJGPP is not suitable for FreeDOS development.

Why Not Add 32-bit Multitasking or Other Advanced Features?

Because we need to ship a 16-bit single-tasking product first. No such free operating system exists. If we keep adding more and more features to what we already have, this will indefinitely prolong and delay completion of FreeDOS.

The goals of the FreeDOS Project are clearly defined. If you have need for a free 32-bit multitasking operating system, we recommend Linux or FreeBSD. Don't worry, you won't hurt our feelings.

Why Use MICRO-C? It Doesn't Support `floats` or `typedefs`!

This is absolutely true. Despite MICRO-C's few, specific limitations, it is an excellent and free product. It is well suited to the task of constructing small command-line-oriented utilities.

Seriously — do you need to pay hundreds of dollars for an advanced commercial C/C++ compiler to develop a program like `xcopy`? By choosing a free compiler to produce a free operating system, not only is the FreeDOS project philosophically consistent with its own goals, but it is easier for new people to participate in development because there is no specific compiler to purchase.

But Wait! Why Is the DOS-C Kernel Written in Borland C?

Development of the DOS-C kernel predates the birth of the FreeDOS Project. Because of the DOS-C kernel's unique position as the foundation of the FreeDOS operating system and because of the time and effort that would be required to port it to MICRO-C or some other compiler, DOS-C development will continue under Borland C.

I Can't Get Some Program to Work.
Is FreeDOS Junk or What?

You could throw up your hands and delete FreeDOS from your hard disk if you run into problems while using it. Or you could be part of the solution by sending us a detailed report of the nature of your particular problem. If we are not aware of problems, we cannot correct them.

What Happened to FreeDOS?
It Seemed Dead for a Long Time.

FreeDOS suffered an extended period of inactivity due to the personal obligations of project coordinator M. "Hannibal" Toal. Now that this period has passed, activity has resumed anew.

What Happened to James Hall,
the Originator of FreeDOS?

James Hall graduated from college in 1995, and due to his personal and career obligations, was no longer able to continue involvement with the FreeDOS Project.

How to Obtain More Information

Where Is the Official FreeDOS Web Page?

The address of the FreeDOS web page is:

`http://sunsite.unc.edu/pub/micro/pc-stuff/freedos/freedos.html`

Where Is the Official FreeDOS FTP Site?

The address of the FreeDOS FTP site is:

```
% ftp sunsite.unc.edu
ftp> cd pub/micro/pc-stuff/freedos
```

Is There a Mailing List for FreeDOS?

The FreeDOS mailing list is maintained by the listserver on `vpro.nl`.

To subscribe to the mailing list:

```
mail free-dos-request@vpro.nl
Subject: subscribe your@email.address
```

Any body text in the message is unnecessary and will not be read.

To cancel a subscription to the mailing list:

```
mail free-dos-request@vpro.nl
Subject: unsubscribe your@email.address
```

You should use the same address that you used when subscribing.

To post a message to the mailing list:

```
mail free-dos@vpro.nl
```

Note that only mail to be posted to the mailing list should be sent to the address `free-dos@vpro.nl`. All new subscription and subscription cancellation requests should be sent to `free-dos-request@vpro.nl`.

Is There a Usenet Newsgroup for FreeDOS?

FreeDOS does not have an official newsgroup. However, official announcements will occur from time to time on the following Usenet newsgroups: `comp.os.msdos.programmer` and `alt.msdos.programmer`.

Who Are the Contact People for FreeDOS?

Project Coordinator	M. "Hannibal" Toal `mtoal@arctic.nsbsd.k12.ak.us`
Kernel Development	Pat Villani `patv@iop.com`
Shell Development	Tim Norman `normat@rpi.edu`
Utility Development:	Bryon Quackenbush `odo@cris.com`
Documentation	BearHeart (Bill Weinman) `bearheart@bearnet.com`
Testing	*This could be you!* `your@name.here`

What Other Documents Exist?

Compatibility File   List of programs known to run under FreeDOS.*

FreeDOS Manifesto   A philosophical statement of Project goals.

Status File   Lists the status of all parts of the Project.

Development Standards   Guidelines for FreeDOS Project development.*

* Denotes documents yet to be written.

FreeDOS is an ongoing project and documentation will be continually updated. Check the FreeDOS home page at:

```
URL http://sunsite.unc.edu/pub/micro/pc-stuff/freedos
```

for the current status of all documentation.

How Can I Participate in FreeDOS Development?

First, congratulations on making it this far! You should now have an idea about what FreeDOS is, what it is not, and where the project is heading. If you would like to become more involved:

- Subscribe to the FreeDOS mailing list. Introduce yourself.

- Download the latest FreeDOS release and try it out.

- Obtain the Compatibility File and test some new unlisted programs.

- Send us your results and opinions!

If you are a crack programmer, or even an aspiring crack programmer:

- Obtain and review the Status File and Development Standards.
- Download MICRO-C and try it out a little bit.
- Identify a task and contact the appropriate development leader.
- Hack some quality code and send it in!

Actually, there is not that much new code to be written. A lot of what needs to be done is porting existing code to MICRO-C, along with extensive testing and error reporting to assist Pat Villani with DOS-C development. FreeDOS is 90 percent completed, but the last 10 percent always seems to take 90 percent of the time.

DOS-C
Source Code

The full source code for this book is now available on the publisher's ftp site at `ftp.mfi.com/pub/rdbooks/FreeDOS.zip`; login as "anonymous" and download the file. Any references to the "companion code diskette" in the book now refer to the code available on the ftp site. For more information, see the FreeDOS home page at

`http://sunsite.unc.edu/pub/micro/pc-stuff/freedos`

or my personal web page at

`http://www.monmouth.com/user_pages/patv`

for updates.

Before You Start

I have made every effort to ensure that the code for this book builds and boots on an IBM-compatible PC. You could simply copy the subdirectory dos-c to your hard drive and build a copy of DOS-C on your computer. However, I'd recommend that you do all your initial testing either on a separate test computer or restrict all operations to floppy disk as you verify its operation. Make multiple copies of the distribution code on disks from which you can work. If you are working with a copy and should accidentally destroy it, you can simply make another copy of the code disk and continue working.

Building DOS-C from the Diskette

Copy the dos-c directory using xcopy, or a similar command, to a disk drive containing a minimum of 5Mb of free space. The distribution is not that large itself, but it requires that amount of space to compile. To build the operating system, a batch file (build.bat) is included in the dos-c directory. On invocation, build.bat proceeds to each subdirectory and builds the boot, kernel, ipl, and utilities. When it completes, the newly created files are in dos-c\dist. In addition, there is a corresponding batch file (clean.bat) to clean up the source directories by removing leftover files, such as *.obj, *.bak, etc.

Directory Structure

```
dos-c          root directory
 +-----dist             holds image of distribution disk
 +-----doc              documentation directory
 +-----hdr              common *.h files
 +-----lib              libm.lib and device.lib
 +-----src              source directories for:
 +--------+-----boot          boot.bin
 +--------+-----command       command.com and help.exe
 +--------+-----drivers       device.lib
 +--------+-----fs            common kernel and ipl fs manager files
 +--------+-----ipl           ipl.sys
 +--------+-----kernel        kernel.exe
 +--------+-----misc          miscellaneous files for kernel and ipl
 +--------+-----tmp
 +--------+-----utils         sys.exe
```

Organization in a Nutshell

Each component or group of utilities is segregated into its own directory. Whenever common files are needed, they are removed and placed in a separate directory. To build that component or utility, a makefile exists in the directory that bears the component's or utility's basename.

Each makefile has at least two targets: production and clean. production builds the expected component or utility and clean cleans up the directory for distribution. The makefile may have at least one additional target that builds the component. I recommend that you examine the component's makefile to better understand how the component is built.

Building a Bootable Floppy

To create a bootable floppy:

1. Format a new floppy. Do not enter a label for the disk; otherwise, the `sys` utility will report an error and abort.

2. Make certain you are in the directory where the DOS-C distribution files are stored. This is normally `dos-c\dist`.

3. Enter the command:

   `sys a:`

   or

   `sys b:`

   to transfer the system files to the diskette.

4. Write protect this disk and boot from it.

Check the distribution disk for additional instructions that are unique to this release.

DOS-C and FreeDOS

Jim Hall began the FreeDOS project when Microsoft began making public statements that it might not support MS-DOS in the future. This statement of Microsoft's marketing plans almost immediately makes every IBM PC, XT, AT, and clones of these machines obsolete. If your machine is not at least an 80386 (or 80486DX2-66, as it turns out), MS-DOS v6.22 is the last new release available to you. Combine this with the software market following Microsoft, and new releases of popular applications are not likely, either. This move condemns you to work with existing applications with the feature/functionality gap growing every day. Jim decided that he would drum up support for an independent MS-DOS operating system that would eventually pick up where Microsoft left off.

At about the same time, I was working with linux and stumbled across the dosemu project. I tried to run my DOS/NT real-time operating system and found that I could boot it under dosemu. I donated my shareware version to the project in the hopes that it would help them while they continued to build DOS features into the emulator. I even considered taking the responsibility of building DOS emulation into dosemu since I was familiar with the interface. Jim Hall was on the dosemu mailing list and became excited when he heard the news.

DOS-C came into being when I was approached by Jim Hall. Jim told me about his effort and I decided not to build the interface into dosemu, but instead to build a stand-alone kernel that could also function as the dosemu kernel. I rewrote the original DOS-C kernel to remove copyrighted material that I did not own because the FreeDOS group wanted to be able to distribute source code under the GNU Public License. The result is what I have described in this book.

However, neither DOS-C nor FreeDOS are static. There are new features and functions being added daily. There are new commands and kernel enhancements, as well as bug fixes. Compatibility problems that may be uncovered during kernel testing are fixed as they arise, introducing yet another reason for continuous updating. Finally, there are plans to move DOS-C and FreeDOS forward with features not found in MS-DOS, so that future versions will have a broader appeal.

With this in mind, I urge you to check the FreeDOS home page at http://sunsite.unc.edu/pub/micro/pc-stuff/freedos or my personal web page at http:/www.monmouth.com/user_pages/patv for updates.

Please Help Us

The FreeDOS project is free software. It relies on the kindness of people like you to pitch in and help create new utilities, document the project, or simply test new software. That's where you come in. If you can, please contact me at patv@iop.com or any of the individuals listed in the FAQ (See Appendix B). The success of the project depends on your support.

Index

Printed in the United States
by Baker & Taylor Publisher Services